The Number System

Landmarks in the Thousands

Grade 4

Also appropriate for Grade 5

Susan Jo Russell
Andee Rubin

Developed at TERC, Cambridge, Massachusetts

Dale Seymour Publications®
White Plains, New York

The *Investigations* curriculum was developed at TERC (formerly Technical Education Research Centers) in collaboration with Kent State University and the State University of New York at Buffalo. The work was supported in part by National Science Foundation Grant No. ESI-9050210. TERC is a nonprofit company working to improve mathematics and science education. TERC is located at 2067 Massachusetts Avenue, Cambridge, MA 02140.

This project was supported, in part, by the

National Science Foundation

Opinions expressed are those of the authors and not necessarily those of the Foundation

Managing Editor: Catherine Anderson
Series Editor: Beverly Cory
Revision Team: Laura Marshall Alavosus, Ellen Harding, Patty Green Holubar, Suzanne Knott, Beverly Hersh Lozoff
ESL Consultant: Nancy Sokol Green
Production/Manufacturing Director: Janet Yearian
Production/Manufacturing Coordinator: Joe Conte
Design Manager: Jeff Kelly
Design: Don Taka
Illustrations: Barbara Epstein-Eagle, Hollis Burkhart
Cover: Bay Graphics
Composition: Archetype Book Composition

This book is published by Dale Seymour Publications®, an imprint of Addison Wesley Longman, Inc.

Dale Seymour Publications
10 Bank Street
White Plains, NY 10602
Customer Service: 1-800-872-1100

Order number DS43893
ISBN 1-57232-746-4
9 10-ML-02

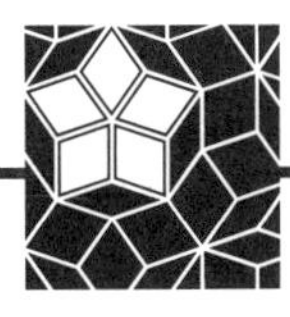
TERC

Principal Investigator **Susan Jo Russell**

Co-Principal Investigator **Cornelia Tierney**

Director of Research and Evaluation **Jan Mokros**

Curriculum Development

Joan Akers
Michael T. Battista
Mary Berle-Carman
Douglas H. Clements
Karen Economopoulos
Ricardo Nemirovsky
Andee Rubin
Susan Jo Russell
Cornelia Tierney
Amy Shulman Weinberg

Evaluation and Assessment

Mary Berle-Carman
Abouali Farmanfarmaian
Jan Mokros
Mark Ogonowski
Amy Shulman Weinberg
Tracey Wright
Lisa Yaffee

Teacher Support

Rebecca B. Corwin
Karen Economopoulos
Tracey Wright
Lisa Yaffee

Technology Development

Michael T. Battista
Douglas H. Clements
Julie Sarama
Andee Rubin

Video Production

David A. Smith

Administration and Production

Amy Catlin
Amy Taber

Cooperating Classrooms for This Unit

Michele de Silva
Angela Philactos
Boston Public Schools
Boston, MA

Kathleen D. O'Connell
Marie Schuler
Arlington Public Schools
Arlington, MA

Consultants and Advisors

Elizabeth Badger
Deborah Lowenberg Ball
Marilyn Burns
Ann Grady
Joanne M. Gurry
James J. Kaput
Steven Leinwand
Mary M. Lindquist
David S. Moore
John Olive
Leslie P. Steffe
Peter Sullivan
Grayson Wheatley
Virginia Woolley
Anne Zarinnia

Graduate Assistants

Joanne Caniglia
Pam DeLong
Carol King
Kent State University

Rosa Gonzalez
Sue McMillen
Julie Sarama
Sudha Swaminathan
State University of New York at Buffalo

Revisions and Home Materials

Cathy Miles Grant
Marlene Kliman
Margaret McGaffigan
Megan Murray
Kim O'Neil
Andee Rubin
Susan Jo Russell
Lisa Seyferth
Myriam Steinback
Judy Storeygard
Anna Suarez
Cornelia Tierney
Carol Walker
Tracey Wright

CONTENTS

TEACHER NOTES

WHERE TO START

The first-time user of *Landmarks in the Thousands* should read the following:

When you next teach this same unit, you can begin to read more of the background. Each time you present the unit, you will learn more about how your students understand the mathematical ideas.

Investigations in Number, Data, and Space® is a K–5 mathematics curriculum with four major goals:

- to offer students meaningful mathematical problems
- to emphasize depth in mathematical thinking rather than superficial exposure to a series of fragmented topics
- to communicate mathematics content and pedagogy to teachers
- to substantially expand the pool of mathematically literate students

The *Investigations* curriculum embodies a new approach based on years of research about how children learn mathematics. Each grade level consists of a set of separate units, each offering 2–8 weeks of work. These units of study are presented through investigations that involve students in the exploration of major mathematical ideas.

Approaching the mathematics content through investigations helps students develop flexibility and confidence in approaching problems, fluency in using mathematical skills and tools to solve problems, and proficiency in evaluating their solutions. Students also build a repertoire of ways to communicate about their mathematical thinking, while their enjoyment and appreciation of mathematics grows.

The investigations are carefully designed to invite all students into mathematics—girls and boys, members of diverse cultural, ethnic, and language groups, and students with different strengths and interests. Problem contexts often call on students to share experiences from their family, culture, or community. The curriculum eliminates barriers—such as work in isolation from peers, or emphasis on speed and memorization—that exclude some students from participating successfully in mathematics. The following aspects of the curriculum ensure that all students are included in significant mathematics learning:

- Students spend time exploring problems in depth.
- They find more than one solution to many of the problems they work on.
- They invent their own strategies and approaches, rather than rely on memorized procedures.
- They choose from a variety of concrete materials and appropriate technology, including calculators, as a natural part of their everyday mathematical work.
- They express their mathematical thinking through drawing, writing, and talking.
- They work in a variety of groupings—as a whole class, individually, in pairs, and in small groups.
- They move around the classroom as they explore the mathematics in their environment and talk with their peers.

While reading and other language activities are typically given a great deal of time and emphasis in elementary classrooms, mathematics often does not get the time it needs. If students are to experience mathematics in depth, they must have enough time to become engaged in real mathematical problems. We believe that a minimum of 5 hours of mathematics classroom time a week—about an hour a day—is critical at the elementary level. The scope and pacing of the *Investigations* curriculum are based on that belief.

We explain more about the pedagogy and principles that underlie these investigations in Teacher Notes throughout the units. For correlations of the curriculum to the NCTM Standards and further help in using this research-based program for teaching mathematics, see the following books, available from Dale Seymour Publications:

- *Implementing the* Investigations in Number, Data, and Space® *Curriculum*
- *Beyond Arithmetic: Changing Mathematics in the Elementary Classroom* by Jan Mokros, Susan Jo Russell, and Karen Economopoulos

This book is one of the curriculum units for *Investigations in Number, Data, and Space.* In addition to providing part of a complete mathematics curriculum for your students, this unit offers information to support your own professional development. You, the teacher, are the person who will make this curriculum come alive in the classroom; the book for each unit is your main support system.

Although the curriculum does not include student textbooks, reproducible sheets for student work are provided in the unit and are also available as Student Activity Booklets. Students work actively with objects and experiences in their own environment and with a variety of manipulative materials and technology, rather than with a book of instruction and problems. We strongly recommend use of the overhead projector as a way to present problems, to focus group discussion, and to help students share ideas and strategies.

Ultimately, every teacher will use these investigations in ways that make sense for his or her particular style, the particular group of students, and the constraints and supports of a particular school environment. Each unit offers information and guidance for a wide variety of situations, drawn from our collaborations with many teachers and students over many years. Our goal in this book is to help you, a professional educator, implement this curriculum in a way that will give all your students access to mathematical power.

Investigation Format

The opening two pages of each investigation help you get ready for the work that follows.

What Happens This gives a synopsis of each session or block of sessions.

Mathematical Emphasis This lists the most important ideas and processes students will encounter in this investigation.

What to Plan Ahead of Time These lists alert you to materials to gather, sheets to duplicate, transparencies to make, and anything else you need to do before starting.

INVESTIGATION 1

Working with 100

What Happens

Session 1: Ways to Count to 100 Students skip count by 2's and by 6's on the 100 chart and discuss patterns they see in the multiples of 2 and 6. Then they find factors of 100, skip count by those factors on miniature 100 charts, and develop conjectures about how to tell which numbers are and aren't factors of 100.

Session 2: 100 in a Box Students make rectangles of different shapes using 100 cubes. They make these "boxes" first with only one layer of cubes (for example, a 10 by 10 array), then with more than one layer (for example, four layers of 5 by 5 squares).

Session 3: Moving Around on the 100 Chart Students are introduced to the idea of "landmarks" in the number system. They explore the difference between various two-digit numbers and 100, one of these important landmarks.

Mathematical Emphasis

- Finding and counting by factors of 100
- Recognizing factor pairs (for example, 4 rows of 25 cubes make 100 and 25 rows of 4 cubes make 100)
- Using landmarks to find differences between numbers under 100 (for example, the difference between 48 and 100 is 52 because from 48 to 50 is 2 and then it's 50 more to 100)
- Making conjectures about factors of 100

2 ▪ *Investigation 1: Working with 100*

INVESTIGATION 1

What to Plan Ahead of Time

Materials

- Overhead projector, transparency pen (Session 1)
- Chart paper: 3 large sheets (Session 1)
- Interlocking cubes: at least 100 per pair (Session 2)
- Chips (counters or centimeter cubes) that fit on 100 chart squares: 3 per student (Session 3)

Other Preparation

- Duplicate student sheets and teaching resources (located at the end of this unit) in the following quantities. If you have Student Activity Booklets, copy only the items marked with an asterisk.

  For Session 1

  Student Sheet 1, Miniature 100 Charts (p. 67): 3–5 per pair

  Student Sheet 2, Factors of 100 (p. 68): 1 per pair

  Family Letter* (p. 66): 1 per student. Remember to sign it before copying.

  Student Sheet 3, More on Factors of 100 (p. 69): 1 per student (homework)

  100 chart (p. 102): 2 per student (class), 1 per student (homework), plus extras*

  For Session 3

  Student Sheet 4, How Far to 100? (p. 70): 1 per student

  Student Sheet 5, Jumping on the 100 Chart (p. 71): 1 per student (homework)

  100 chart (p. 102): 1 per student (class), 1 per student (homework)

- Students will use the 100 chart as a standard material throughout the unit. Make sure there is always a supply of these readily available for all students.
- Make three overhead transparencies of the 100 chart (p. 102). (Session 1)
- If you plan to provide folders in which students will save their work for the entire unit, prepare these for distribution during Session 1.

Investigation 1: Working with 100 ▪ 3

Sessions Within an investigation, the activities are organized by class session, a session being at least a one-hour math class. Sessions are numbered consecutively through an investigation. Often several sessions are grouped together, presenting a block of activities with a single major focus.

When you find a block of sessions presented together—for example, Sessions 1, 2, and 3—read through the entire block first to understand the overall flow and sequence of the activities. Make some preliminary decisions about how you will divide the activities into three sessions for your class, based on what you know about your students. You may need to modify your initial plans as you progress through the activities, and you may want to make notes in the margins of the pages as reminders for the next time you use the unit.

Be sure to read the Session Follow-Up section at the end of the session block to see what homework assignments and extensions are suggested as you make your initial plans.

While you may be used to a curriculum that tells you exactly what each class session should cover, we have found that the teacher is in a better position to make these decisions. Each unit is flexible and may be handled somewhat differently by every teacher. Although we provide guidance for how many sessions a particular group of activities is likely to need, we want you to be active in determining an appropriate pace and the best transition points for your class. It is not unusual for a teacher to spend more or less time than is proposed for the activities.

Ten-Minute Math At the beginning of some sessions, you will find Ten-Minute Math activities. These are designed to be used in tandem with the investigations, but not during the math hour. Rather, we hope you will do them whenever you have a spare 10 minutes—maybe before lunch or recess, or at the end of the day.

Ten-Minute Math offers practice in key concepts, but not always those being covered in the unit. For example, in a unit on using data, Ten-Minute Math must revisit geometric activities done earlier in the year. Complete directions for the suggested activities are included at the end of each unit.

Sessions 2, 3, and 4

Using Landmarks to Add and Subtract

What Happens

Students develop strategies for solving sets of related addition and subtraction problems, using a result from one problem (for example, 100 – 20) to solve an unknown one (for example, 100 – 22). During Choice Time, students choose from several activities: finding factors of 300, making boxes for 200 cubes, doing related problem sets involving numbers to 300, and playing a game of Bingo on the 101 to 200 chart. Their work focuses on:

- adding and subtracting numbers in the 100's
- using landmarks and skip counting to solve problems
- finding factors of 200 and 300
- adding multiples of 10 to and subtracting multiples of 10 from numbers on the 300 chart

Materials

- Interlocking cubes, 100 charts, and 300 charts, and chips for the charts remain available
- Calculators (1 per pair)
- Student Sheet 7 (6 of each page, cut into cards)
- Student Sheet 8 (1 per student)
- Student Sheet 9 (1 per student, homework)
- Student Sheet 10 (1 per student, homework)
- 300 chart (2 per student, homework)
- One-centimeter graph paper (1–2 pieces per student)
- 101 to 200 Bingo Board (2 per student, class; 2 per student, homework)
- How to Play 101 to 200 Bingo (1 per student, homework, optional)
- Numeral Cards (1 deck per group, class; 1 deck per student, homework)
- Tens Cards (1 deck per group, class; 1 deck per student, homework)
- Envelopes (1 per student)
- Pencils, crayons, or markers

Sessions 2, 3, and 4: Using Landmarks to Add and Subtract ■ **23**

Activities The activities include pair and small-group work, individual tasks, and whole-class discussions. In any case, students are seated together, talking and sharing ideas during all work times. Students most often work cooperatively, although each student may record work individually.

Choice Time In most units, some sessions are structured with activity choices. In these cases, students may work simultaneously on different activities focused on the same mathematical ideas. Students choose which activities they want to do, and they cycle through them.

You will need to decide how to set up and introduce these activities and how to let students make their choices. Some teachers present them as station activities, in different parts of the room. Some list the choices on the board as reminders or have students keep their own lists.

Tips for the Linguistically Diverse Classroom At strategic points in each unit, you will find concrete suggestions for simple modifications of the teaching strategies to encourage the participation of all students. Many of these tips offer alternative ways to elicit critical thinking from students at varying levels of English proficiency, as well as from other students who find it difficult to verbalize their thinking.

The tips are supported by suggestions for specific vocabulary work to help ensure that all students can participate fully in the investigations. The Preview for the Linguistically Diverse Classroom lists important words that are assumed as part of the working vocabulary of the unit. Second-language learners will need to become familiar with these words in order to understand the problems and activities they will be doing. These terms can be incorporated into students' second-language work before or during the unit. Activities that can be used to present the words are found in the appendix, Vocabulary Support for Second-Language Learners. In addition, ideas for making connections to students' languages and cultures, included on the Preview page, help the class explore the unit's concepts from a multicultural perspective.

Session Follow-Up: Homework In *Investigations,* homework is an extension of classroom work. Sometimes it offers review and practice of work done in class, sometimes preparation for upcoming activities, and sometimes numerical practice that revisits work in earlier units. Homework plays a role both in supporting students' learning and in helping inform families about the ways in which students in this curriculum work with mathematical ideas.

Depending on your school's homework policies and your own judgment, you may want to assign more homework than is suggested in the units. For this purpose you might use the practice pages, included as blackline masters at the end of this unit, to give students additional work with numbers.

For some homework assignments, you will want to adapt the activity to meet the needs of a variety of students in your class: those with special needs, those ready for more challenge, and second-language learners. You might change the numbers in a problem, make the activity more or less complex, or go through a sample activity with those who need extra help. You can modify any student sheet for either homework or class use. In particular, making numbers in a problem smaller or larger can make the same basic activity appropriate for a wider range of students.

Another issue to consider is how to handle the homework that students bring back to class—how to recognize the work they have done at home without spending too much time on it. Some teachers hold a short group discussion of different approaches to the assignment; others ask students to share and discuss their work with a neighbor; still others post the homework around the room and give students time to tour it briefly. If you want to keep track of homework students bring in, be sure it ends up in a designated place.

Session Follow-Up: Extensions Sometimes in Session Follow-Up, you will find suggested extension activities. These are opportunities for some or all students to explore a topic in greater depth or in a different context. They are not designed for "fast" students; mathematics is a multifaceted discipline, and different students will want to go further in different investigations. Look for and encourage the sparks of interest and enthusiasm you see in your students, and use the extensions to help them pursue these interests.

Excursions Some of the *Investigations* units include excursions—blocks of activities that could be omitted without harming the integrity of the unit. This is one way of dealing with the great depth and variety of elementary mathematics—much more than a class has time to explore in any one year. Excursions give you the flexibility to make different choices from year to year, doing the excursion in one unit this time, and next year trying another excursion.

Materials

A complete list of the materials needed for teaching this unit follows the unit overview. Some of these materials are available in kits for the *Investigations* curriculum. Individual items can also be purchased from school supply dealers.

Classroom Materials In an active mathematics classroom, certain basic materials should be available at all times: interlocking cubes, pencils, unlined paper, graph paper, calculators, things to count with, and measuring tools. Some activities in this curriculum require scissors and glue sticks or tape. Stick-on notes and large paper are also useful materials throughout.

So that students can independently get what they need at any time, they should know where these materials are kept, how they are stored, and how they are to be returned to the storage area. For example, interlocking cubes are best stored in towers of ten; then, whatever the activity, they should be returned to storage in groups of ten at the end of the hour. You'll find that establishing such routines at the beginning of the year is well worth the time and effort.

Student Sheets and Teaching Resources Student recording sheets and other teaching tools needed for both class and homework are provided as reproducible blackline masters at the end of each unit. We think it's important that students find their own ways of organizing and recording their work. They need to learn how to explain their thinking with both drawings and written words, and how to organize their results so someone else can understand them. For this reason, we deliberately do not provide student sheets for every activity. Regardless of the form in which students do their work, we recommend that they keep their work in a mathematics folder, notebook, or journal so that it is always available to them for reference.

Name Date

Student Sheet 16

What's Your Estimate?

1. Which container are you using?

2. First, just look at the container. How many do you think there are in the container?

   We think the total number is closest to (circle one of these):

   100 200 300 400 500 600 700 800 900 1000

3. Now try to figure out a better estimate without counting one by one. When you are finished, write down your new estimate and what method you used.

   Here is what we did:

   Now we think the total number is closest to (circle one of these):

   100 200 300 400 500 600 700 800 900 1000

 94 *Investigation 3 • Sessions 3–5* *Landmarks in the Thousands*

Student Activity Booklets These booklets contain all the sheets each student will need for individual work, freeing you from extensive copying (although you may need or want to copy the occasional teaching resource on transparency film or card stock, or make extra copies of a student sheet).

Calculators and Computers Calculators are used throughout Investigations. Many of the unity recommend that you have at least one calculator for each pair. You will find calculator activities, plus Teacher Notes discussing this important mathematical tool, in an early unit at each grade level. It is assumed that calculators will be readily available for student use.

Computer activities are offered at all grade levels. How you use the computer activities depends on the number of computers you have available. Technology in the Curriculum discusses ways to incorporate the use of calculators and computers into classroom activities.

Children's Literature Each unit offers a list of related children's literature that can be used to support the mathematical ideas in the unit. Sometimes an activity is based on a specific children's book, with suggestions for substitutions where practical. While such activities can be adapted and taught without the book, the literature offers a rich introduction and should be used whenever possible.

***Investigations* at Home** It is a good idea to make your policy on homework explicit to both students and their families when you begin teaching with *Investigations*. How frequently will you be assigning homework? When do you expect homework to be completed and brought back to school? What are your goals in assigning homework? How independent should families expect their children to be? What should the parent's or guardian's role be? The more explicit you can be about your expectations, the better the homework experience will be for everyone.

Investigations at Home (a booklet available separately for each unit, to send home with students) gives you a way to communicate with families about the work students are doing in class. This booklet includes a brief description of every session, a list of the mathematics content emphasized in each investigation, and a discussion of each homework assignment to help families more effectively support their children. Whether or not you are using the *Investigations* at Home booklets, we expect you to make your own choices about homework assignments. Feel free to omit any and to add extra ones you think are appropriate.

Family Letter A letter that you can send home to students' families is included with the blackline masters for each unit. Families need to be informed about the mathematics work in your classroom; they should be encouraged to participate in and support their children's work. A reminder to send home the letter for each unit appears in one of the early investigations. These letters are also available separately in Spanish, Vietnamese, Cantonese, Hmong, and Cambodian.

Help for You, the Teacher

Because we believe strongly that a new curriculum must help teachers think in new ways about mathematics and about their students' mathematical thinking processes, we have included a great deal of material to help you learn more about both.

About the Mathematics in This Unit This introductory section summarizes the critical information about the mathematics you will be teaching. It describes the unit's central mathematical ideas and the ways students will encounter them through the unit's activities.

About the Assessment in This Unit This introductory section highlights Teacher Checkpoints and assessment activities contained in the unit. It offers questions to stimulate your assessment as you observe the development of students' mathematical thinking and learning.

Teacher Notes These reference notes provide practical information about the mathematics you are teaching and about our experience with how students learn. Many of the notes were written in response to actual questions from teachers or to discuss important things we saw happening in the field-test classrooms. Some teachers like to read them all before starting the unit, then review them as they come up in particular investigations.

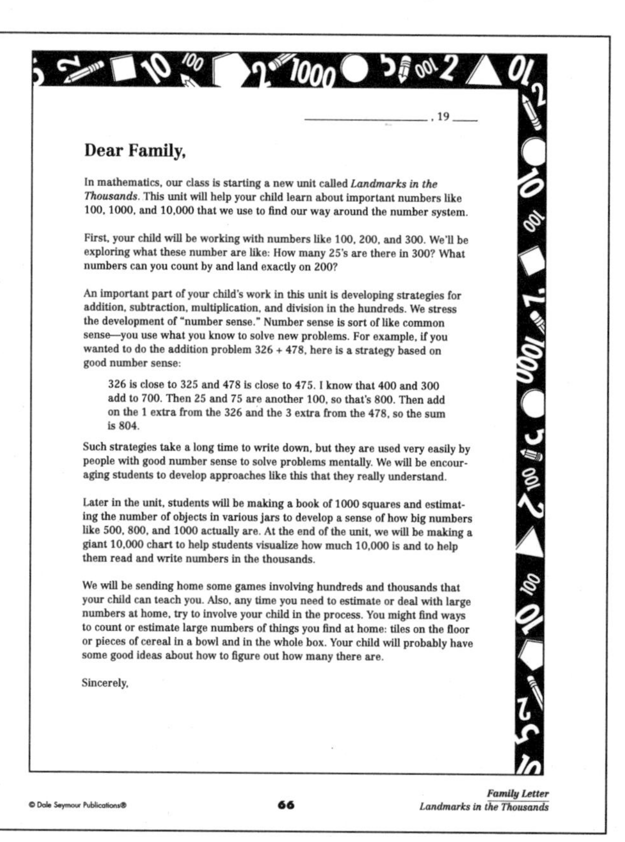

____________, 19____

Dear Family,

In mathematics, our class is starting a new unit called *Landmarks in the Thousands*. This unit will help your child learn about important numbers like 100, 1000, and 10,000 that we use to find our way around the number system.

First, your child will be working with numbers like 100, 200, and 300. We'll be exploring what these number are like: How many 25's are there in 300? What numbers can you count by and land exactly on 200?

An important part of your child's work in this unit is developing strategies for addition, subtraction, multiplication, and division in the hundreds. We stress the development of "number sense." Number sense is sort of like common sense—you use what you know to solve new problems. For example, if you wanted to do the addition problem 326 + 478, here is a strategy based on good number sense:

> 326 is close to 325 and 478 is close to 475. I know that 400 and 300 add to 700. Then 25 and 75 are another 100, so that's 800. Then add on the 1 extra from the 326 and the 3 extra from the 478, so the sum is 804.

Such strategies take a long time to write down, but they are used very easily by people with good number sense to solve problems mentally. We will be encouraging students to develop approaches like this that they really understand.

Later in the unit, students will be making a book of 1000 squares and estimating the number of objects in various jars to develop a sense of how big numbers like 500, 800, and 1000 actually are. At the end of the unit, we will be making a giant 10,000 chart to help students visualize how much 10,000 is and to help them read and write numbers in the thousands.

We will be sending home some games involving hundreds and thousands that your child can teach you. Also, any time you need to estimate or deal with large numbers at home, try to involve your child in the process. You might find ways to count or estimate large numbers of things you find at home: tiles on the floor or pieces of cereal in a bowl and in the whole box. Your child will probably have some good ideas about how to figure out how many there are.

Sincerely,

66
Family Letter
Landmarks in the Thousands

Dialogue Boxes Sample dialogues demonstrate how students typically express their mathematical ideas, what issues and confusions arise in their thinking, and how some teachers have guided class discussions.

These dialogues are based on the extensive classroom testing of this curriculum; many are word-for-word transcriptions of recorded class discussions. They are not always easy reading; sometimes it may take some effort to unravel what the students are trying to say. But this is the value of these dialogues; they offer good clues to how your students may develop and express their approaches and strategies, helping you prepare for your own class discussions.

Where to Start You may not have time to read everything the first time you use this unit. As a first-time user, you will likely focus on understanding the activities and working them out with your students. Read completely through all the activities before starting to present them. Also read those sections listed in the Contents under the heading Where to Start.

Teacher Note **_Boxes for 100_**

While making their boxes of 100, students may bring up some important mathematical ideas:

- The dimensions of their boxes, both two-dimensional and three-dimensional, are the same numbers they found when they skip counted on the 100 chart with numbers that land exactly on 100—the *factors* of 100. Here are questions to pose: "Why do you think the same numbers appear in both situations? Here's a box of 2 rows of 50. What does 2 rows of 50 have to do with what you can do on the 100 chart? On the 100 chart, you found out you could skip count by 4's and land on 100, so 4 is a factor of 100. Is there anything the same about skip counting by 4 and any of the boxes you made?"
- Factors come in pairs. If you find that 4 is a factor of 100, then there is another number which, when multiplied by 4, gives you 100 (in this case, 25). This number is also a factor of 100. Here are questions to ask: "Is this always true? Do factors always come in pairs? Why do they? If 4 × 25 is equal to 100, does 25 × 4 have to be equal to 100? Is this true for other numbers? What about 10? Is there another factor of 100 it is paired with?"
- Five rows of 20 and 20 rows of 5 make the same-shape box. Here is a question to pose: "Should we consider these boxes to be different or the same?" There is not one answer to this question: Both sides of the question can be defended. For example, students might contend that because you can simply rotate the box and make it look exactly like the other box, they are the same—it wouldn't make any difference to the caramel company which way people held the box. On the other hand, someone might say that if you spread the rows apart, 5 rows of 20 (a few very long rows) looks very different from 20 rows of 5 (a lot of short rows). If the caramel company puts dividers between the rows, five 20's and twenty 5's might look different to the eye.

The same issue will come up when students make multilayered boxes: "Is a box made of 2 layers that are each 5 by 10 the same as a box made of 5 layers that are each 10 by 2?" Again, there are arguments for both points of view, although the argument for considering them as different, at least in this context of boxes for caramels, may be even stronger than for one-layer boxes. After all, a rectangular solid with a 5 by 10 rectangle as its base really does look different from one with a 2 by 10 rectangle as its base.

If boxes with the same dimensions are considered the same no matter which part is on the bottom, there are three different multilayer boxes: 2 by 2 by 25, 2 by 5 by 10, and 4 by 5 by 5. If unique boxes are determined by the rectangle on the bottom, there are seven different boxes with the bottoms: 2 by 2, 2 by 25, 2 by 5, 2 by 10, 5 by 10, 4 by 5, and 5 by 5.

14 ▪ *Investigation 1: Working with 100*

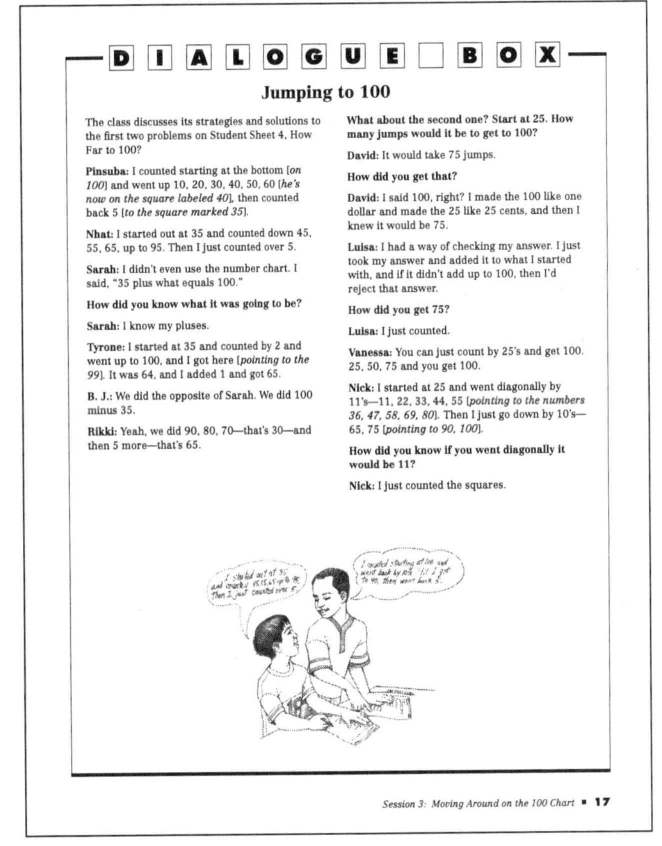

DIALOGUE BOX

Jumping to 100

The class discusses its strategies and solutions to the first two problems on Student Sheet 4, How Far to 100?

Pinsuba: I counted starting at the bottom [*on 100*] and went up 10, 20, 30, 40, 50, 60 [*he's now on the square labeled 40*], then counted back 5 [*to the square marked 35*].

Nhat: I started out at 35 and counted down 45, 55, 65, up to 95. Then I just counted over 5.

Sarah: I didn't even use the number chart. I said, "35 plus what equals 100."

How did you know what it was going to be?

Sarah: I know my pluses.

Tyrone: I started at 35 and counted by 2 and went up to 100, and I got here [*pointing to the 99*]. It was 64, and I added 1 and got 65.

B. J.: We did the opposite of Sarah. We did 100 minus 35.

Rikki: Yeah, we did 90, 80, 70—that's 30—and then 5 more—that's 65.

What about the second one? Start at 25. How many jumps would it be to get to 100?

David: It would take 75 jumps.

How did you get that?

David: I said 100, right? I made the 100 like one dollar and made the 25 like 25 cents, and then I knew it would be 75.

Luisa: I had a way of checking my answer. I just took my answer and added it to what I started with, and if it didn't add up to 100, then I'd reject that answer.

How did you get 75?

Luisa: I just counted.

Vanessa: You can just count by 25's and get 100. 25, 50, 75 and you get 100.

Nick: I started at 25 and went diagonally by 11's—11, 22, 33, 44, 55 [*pointing to the numbers 36, 47, 58, 69, 80*]. Then I just go down by 10's—65, 75 [*pointing to 90, 100*].

How did you know if you went diagonally it would be 11?

Nick: I just counted the squares.

Session 3: Moving Around on the 100 Chart ▪ 17

The *Investigations* curriculum incorporates the use of two forms of technology in the classroom: calculators and computers. Calculators are assumed to be standard classroom materials, available for student use in any unit. Computers are explicitly linked to one or more units at each grade level; they are used with the unit on 2-D geometry at each grade, as well as with some of the units on measuring, data, and changes.

Using Calculators

In this curriculum, calculators are considered tools for doing mathematics, similar to pattern blocks or interlocking cubes. Just as with other tools, students must learn both *how* to use calculators correctly and *when* they are appropriate to use. This knowledge is crucial for daily life, as calculators are now a standard way of handling numerical operations, both at work and at home.

Using a calculator correctly is not a simple task; it depends on a good knowledge of the four operations and of the number system, so that students can select suitable calculations and also determine what a reasonable result would be. These skills are the basis of any work with numbers, whether or not a calculator is involved.

Unfortunately, calculators are often seen as tools to check computations with, as if other methods are somehow more fallible. Students need to understand that any computational method can be used to check any other; it's just as easy to make a mistake on the calculator as it is to make a mistake on paper or with mental arithmetic. Throughout this curriculum, we encourage students to solve computation problems in more than one way in order to double-check their accuracy. We present mental arithmetic, paper-and-pencil computation, and calculators as three possible approaches.

In this curriculum we also recognize that, despite their importance, calculators are not always appropriate in mathematics instruction. Like any tools, calculators are useful for some tasks but not for others. You will need to make decisions about when to allow students access to calculators and when to ask that they solve problems without them so that they can concentrate on other tools and skills. At times when calculators are or are not appropriate for a particular activity, we make specific recommendations. Help your students develop their own sense of which problems they can tackle with their own reasoning and which ones might be better solved with a combination of their own reasoning and the calculator.

Managing calculators in your classroom so that they are a tool, and not a distraction, requires some planning. When calculators are first introduced, students often want to use them for everything, even problems that can be solved quite simply by other methods. However, once the n ovelty wears off, students are just as interested in developing their own strategies, especially when these strategies are emphasized and valued in the classroom. Over time, students will come to recognize the ease and value of solving problems mentally, with paper and pencil, or with manipulatives, while also understanding the power of the calculator to facilitate work with larger numbers.

Experience shows that if calculators are available only occasionally, students become excited and distracted when they are permitted to use them. They focus on the tool rather than on the mathematics. In order to learn when calculators are appropriate and when they are not, students must have easy access to them and use them routinely in their work.

If you have a calculator for each student, and if you think your students can accept the responsibility, you might allow them to keep their calculators with the rest of their individual materials, at least for the first few weeks of school. Alternatively, you might store them in boxes on a shelf, number each calculator, and assign a corresponding number to each student. This system can give students a sense of ownership while also helping you keep track of the calculators.

Using Computers

Students can use computers to approach and visualize mathematical situations in new ways. The computer allows students to construct and manipulate geometric shapes, see objects move according to rules they specify, and turn, flip, and repeat a pattern.

This curriculum calls for computers in units where they are a particularly effective tool for learning mathematics content. One unit on 2-D geometry at each of the grades 3–5 includes a core of activities that rely on access to computers, either in the classroom or in a lab. Other units on geometry, measuring, data, and changes include computer activities, but can be taught without them. In these units, however, students' experience is greatly enhanced by computer use.

The following list outlines the recommended use of computers in this curriculum:

Kindergarten
Unit: *Making Shapes and Building Blocks* (Exploring Geometry)
Software: *Shapes*
Source: provided with the unit

Grade 1
Unit: *Survey Questions and Secret Rules* (Collecting and Sorting Data)
Software: *Tabletop, Jr.*
Source: Broderbund

Unit: *Quilt Squares and Block Towns* (2-D and 3-D Geometry)
Software: *Shapes*
Source: provided with the unit

Grade 2
Unit: *Mathematical Thinking at Grade 2* (Introduction)
Software: *Shapes*
Source: provided with the unit

Unit: *Shapes, Halves, and Symmetry* (Geometry and Fractions)
Software: *Shapes*
Source: provided with the unit

Unit: *How Long? How Far?* (Measuring)
Software: *Geo-Logo*
Source: provided with the unit

Grade 3
Unit: *Flips, Turns, and Area* (2-D Geometry)
Software: *Tumbling Tetrominoes*
Source: provided with the unit

Unit: *Turtle Paths* (2-D Geometry)
Software: *Geo-Logo*
Source: provided with the unit

Grade 4
Unit: *Sunken Ships and Grid Patterns* (2-D Geometry)
Software: *Geo-Logo*
Source: provided with the unit

Grade 5
Unit: *Picturing Polygons* (2-D Geometry)
Software: *Geo-Logo*
Source: provided with the unit

Unit: *Patterns of Change* (Tables and Graphs)
Software: *Trips*
Source: provided with the unit

Unit: *Data: Kids, Cats, and Ads* (Statistics)
Software: *Tabletop, Sr.*
Source: Broderbund

The software provided with the *Investigations* units uses the power of the computer to help students explore mathematical ideas and relationships that cannot be explored in the same way with physical materials. With the *Shapes* (grades 1–2) and *Tumbling Tetrominoes* (grade 3) software, students explore symmetry, pattern, rotation and reflection, area, and characteristics of 2-D shapes. With the *Geo-Logo* software (grades 2–5), students investigate rotations and reflections, coordinate geometry, the properties of 2-D shapes, and angles. The *Trips* software (grade 5) is a mathematical exploration of motion in which students run experiments and interpret data presented in graphs and tables.

We suggest that students work in pairs on the computer; this not only maximizes computer resources but also encourages students to consult, monitor, and teach each other. Generally, more than two students at one computer find it difficult to share. Managing access to computers is an issue for every classroom. The curriculum gives you explicit support for setting up a system. The units are structured on the assumption that you have enough computers for half your students to work on the machines in pairs at one time. If you do not have access to that many computers, suggestions are made for structuring class time to use the unit with fewer than five.

Assessment plays a critical role in teaching and learning, and it is an integral part of the *Investigations* curriculum. For a teacher using these units, assessment is an ongoing process. You observe students' discussions and explanations of their strategies on a daily basis and examine their work as it evolves. While students are busy recording and representing their work, working on projects, sharing with partners, and playing mathematical games, you have many opportunities to observe their mathematical thinking. What you learn through observation guides your decisions about how to proceed. In any of the units, you will repeatedly consider questions like these:

- Do students come up with their own strategies for solving problems, or do they expect others to tell them what to do? What do their strategies reveal about their mathematical understanding?
- Do students understand that there are different strategies for solving problems? Do they articulate their strategies and try to understand other students' strategies?
- How effectively do students use materials as tools to help with their mathematical work?
- Do students have effective ideas for keeping track of and recording their work? Do keeping track of and recording their work seem difficult for them?

You will need to develop a comfortable and efficient system for recording and keeping track of your observations. Some teachers keep a clipboard handy and jot notes on a class list or on adhesive labels that are later transferred to student files. Others keep loose-leaf notebooks with a page for each student and make weekly notes about what they have observed in class.

Assessment Tools in the Unit

With the activities in each unit, you will find questions to guide your thinking while observing the students at work. You will also find two built-in assessment tools: Teacher Checkpoints and embedded Assessment activities.

Teacher Checkpoints The designated Teacher Checkpoints in each unit offer a time to "check in" with individual students, watch them at work, and ask questions that illuminate how they are thinking.

At first it may be hard to know what to look for, hard to know what kinds of questions to ask. Students may be reluctant to talk; they may not be accustomed to having the teacher ask them about their work, or they may not know how to explain their thinking. Two important ingredients of this process are asking students open-ended questions about their work and showing genuine interest in how they are approaching the task. When students see that you are interested in their thinking and are counting on them to come up with their own ways of solving problems, they may surprise you with the depth of their understanding.

Teacher Checkpoints also give you the chance to pause in the teaching sequence and reflect on how your class is doing overall. Think about whether you need to adjust your pacing: Are most students fluent with strategies for solving a particular kind of problem? Are they just starting to formulate good strategies? Or are they still struggling with how to start? Depending on what you see as the students work, you may want to spend more time on similar problems, change some of the problems to use smaller numbers, move quickly to more challenging material, modify subsequent activities for some students, work on particular ideas with a small group, or pair students who have good strategies with those who are having more difficulty.

Embedded Assessment Activities Assessment activities embedded in each unit will help you examine specific pieces of student work, figure out what they mean, and provide feedback. From the students' point of view, these assessment activities are no different from any others. Each is a learning experience in and of itself, as well as an opportunity for you to gather evidence about students' mathematical understanding.

The embedded assessment activities sometimes involve writing and reflecting; at other times, a discussion or brief interaction between student and teacher; and in still other instances, the creation and explanation of a product. In most cases, the assessments require that students *show* what they did, *write* or *talk* about it, or do both. Having to explain how they worked through a problem helps students be more focused and clear in their mathematical thinking. It also helps them realize that doing mathematics is a process that may involve tentative starts, revising one's approach, taking different paths, and working through ideas.

Teachers often find the hardest part of assessment to be interpreting their students' work. We provide guidelines to help with that interpretation. If you have used a process approach to teaching writing, the assessment in *Investigations* will seem familiar. For many of the assessment activities, a Teacher Note provides examples of student work and a commentary on what it indicates about student thinking.

Documentation of Student Growth

To form an overall picture of mathematical progress, it is important to document each student's work. Many teachers have students keep their work in folders, notebooks, or journals, and some like to have students summarize their learning in journals at the end of each unit. It's important to document students' progress, and we recommend that you keep a portfolio of selected work for each student, unit by unit, for the entire year. The final activity in each *Investigations* unit, called Choosing Student Work to Save, helps you and the students select representative samples for a record of their work.

This kind of regular documentation helps you synthesize information about each student as a mathematical learner. From different pieces of evidence, you can put together the big picture. This synthesis will be invaluable in thinking about where to go next with a particular child, deciding where more work is needed, or explaining to parents (or other teachers) how a child is doing.

If you use portfolios, you need to collect a good balance of work, yet avoid being swamped with an overwhelming amount of paper. Following are some tips for effective portfolios:

- Collect a representative sample of work, including some pieces that students themselves select for inclusion in the portfolio. There should be just a few pieces for each unit, showing different kinds of work—some assignments that involve writing as well as some that do not.
- If students do not date their work, do so yourself so that you can reconstruct the order in which pieces were done.
- Include your reflections on the work. When you are looking back over the whole year, such comments are reminders of what seemed especially interesting about a particular piece; they can also be helpful to other teachers and to parents. Older students should be encouraged to write their own reflections about their work.

Assessment Overview

There are two places to turn for a preview of the assessment opportunities in each *Investigations* unit. The Assessment Resources column in the unit Overview Chart identifies the Teacher Checkpoints and Assessment activities embedded in each investigation, guidelines for observing the students that appear within classroom activities, and any Teacher Notes and Dialogue Boxes that explain what to look for and what types of student responses you might expect to see in your classroom. Additionally, the section About the Assessment in This Unit gives you a detailed list of questions for each investigation, keyed to the mathematical emphases, to help you observe student growth.

Depending on your situation, you may want to provide additional assessment opportunities. Most of the investigations lend themselves to more frequent assessment, simply by having students do more writing and recording while they are working.

Landmarks in the Thousands

Content of This Unit Students explore the structure of our number system through activities involving hundreds and thousands. They explore factors of 100 and 1000, important landmarks in our number system, and use this experience to work with multiples of 100 and 1000. They use these landmarks to solve addition and subtraction problems in the hundreds. They also play games and solve problems that focus on what happens when we add or subtract multiples of 10 and 100, another critical feature of the base ten system. Through estimating quantities of objects and making a 1000 book and a 10,000 class chart, they begin to develop a sense of the relative sizes of 100, 1000, and 10,000.

Connections with Other Units If you are doing the full-year *Investigations* curriculum in the suggested sequence for grade 4, this is the fourth of eleven units. In the introductory unit, *Mathematical Thinking at Grade 4*, students began to develop strategies to solve addition and subtraction problems. Your class will also already have completed the Grade 4 Multiplication and Division unit *Arrays and Shares*, in which your students gained experience with finding factors of numbers and solving related problem sets dealing with multiplication and division. If students have not had comparable experience, you may need to spend more time skip counting and finding factors in Investigation 1.

If your school is not using the full-year curriculum, this unit can also be used successfully at grade 5. The work in this unit is continued and extended in the Grade 4 Multiplication and Division unit *Packages and Groups*.

Investigations Curriculum ■ Suggested Grade 4 Sequence

Mathematical Thinking at Grade 4 (Introduction)

Arrays and Shares (Multiplication and Division)

Seeing Solids and Silhouettes (3-D Geometry)

▶ *Landmarks in the Thousands* (The Number System)

Different Shapes, Equal Pieces (Fractions and Area)

The Shape of the Data (Statistics)

Money, Miles, and Large Numbers (Addition and Subtraction)

Changes Over Time (Graphs)

Packages and Groups (Multiplication and Division)

Sunken Ships and Grid Patterns (2-D Geometry)

Three out of Four Like Spaghetti (Data and Fractions)

Investigation 1 ■ Working with 100		
Class Sessions	**Activities**	**Pacing**
Session 1 (p. 4) WAYS TO COUNT TO 100	Skip Counting on the 100 Chart by 2's and 6's Highlighting 100 Charts What Are All the Factors of 100? Discussion: Do We Have All the Factors of 100? Homework: More on Factors of 100 Extension: A Book of 100 Charts Extension: Finding Factors	minimum 1 hr
Session 2 (p. 11) 100 IN A BOX	Building with 100 Cubes	minimum 1 hr
Session 3 (p. 15) MOVING AROUND ON THE 100 CHART	Teacher Checkpoint: Jumping to 100 Homework: Jumping on the 100 Chart	minimum 1 hr

Mathematical Emphasis

- Finding and counting by factors of 100
- Recognizing factor pairs (for example, 4 rows of 25 cubes make 100 and 25 rows of 4 cubes make 100)
- Using landmarks to find differences between numbers under 100 (for example, the difference between 48 and 100 is 52 because from 48 to 50 is 2 and then it is 50 more to 100)
- Making conjectures about factors of 100

Assessment Resources

Counting by 2's and 6's (Dialogue Box, p. 9)

Introducing Mathematical Vocabulary (Teacher Note, p. 10)

Students' Difficulties Making Arrays (Teacher Note, p. 13)

Boxes for 100 (Teacher Note, p. 14)

Teacher Checkpoint: Jumping to 100 (p. 15)

Jumping to 100 (Dialogue Box, p. 17)

Materials

Overhead projector, transparencies, and pens

Chart paper

Interlocking cubes

Chips, counters, or centimeter cubes

Family letter

Student Sheets 1–5

Teaching resource sheet

Investigation 2 ▪ Exploring Multiples of 100

Class Sessions	Activities	Pacing
Session 1 (p. 20) FACTORS OF 100, 200, AND 300	Skip Counting on the 300 Chart Factors of 200, 300, and Up!	minimum 1 hr
Sessions 2, 3, and 4 (p. 23) USING LANDMARKS TO ADD AND SUBTRACT	Related Problem Sets Teacher Checkpoint: Choice Time: Working with 200 and 300 Discussing Related Problem Sets Homework: More Related Problem Sets Homework: Froggy Races 1 Homework: 101 to 200 Bingo Extension: Multilayer Boxes for 200	minimum 3 hr
Session 5 (p. 34) SOLVING PROBLEMS IN THE HUNDREDS	Assessment: Problems with Landmarks in the Hundreds Homework: Another Problem in Two Ways	minimum 1 hr

Ten-Minute Math ▪ What Is Likely?

Mathematical Emphasis

- Using knowledge about the factors of 100 to explore multiples of 100 (for example, if there are four 25's in 100, then there are eight in 200 and twelve in 300)
- Relating knowledge of factors to division situations and to standard division notation
- Adding and subtracting multiples of 10 to numbers in the hundreds
- Solving addition and subtraction problems by reasoning from known relationships
- Communicating strategies orally and on paper through use of words, pictures, and numbers

Assessment Resources

Teacher Checkpoint: Boxes for 200 (p. 27)

Related Problem Sets (Teacher Note, p. 30)

Students' Strategies for Solving Related Problem Sets (Dialogue Box, p. 32)

Is It 1 More or 1 Less? (Dialogue Box, p. 33)

Assessment: Problems with Landmarks in the Hundreds (p. 34)

Materials

Chips, counters, or centimeter cubes

Crayons or markers

Interlocking cubes

Calculators

Envelopes

Student Sheets 6–12

Teaching resource sheets

Investigation 3 ■ How Much Is 1000?

Class Sessions	Activities	Pacing
Session 1 (p. 38) NUMBERS TO 1000	Making a 1000 Book Teacher Checkpoint: Find the Numbers Homework: Numbers in My 1000 Book	minimum 1 hr
Session 2 (p. 41) MOVING AROUND IN THE 1000 BOOK	Factors of 1000: Counting with Calculators Counting to 1000 How Far to 1000?	minimum 1 hr
Sessions 3, 4, and 5 (p. 43) ESTIMATING, ADDING, AND SUBTRACTING TO 1000	Choice Time: Working with Hundreds Looking at Our Estimates Assessment: Make Your Own Related Problem Set Homework: Froggy Races 2 Homework: Close to 1000 Extension: Finding 1000 Things Extension: Counting Around the Class: What Would Get Us Closest to 1000?	minimum 3 hr

Ten-Minute Math ■ Counting Around the Class

Mathematical Emphasis

- Reading and writing numbers to 1000
- Locating numbers in sequence to 1000
- Getting a sense of the magnitude of multiples of 100 up to 1000
- Identifying and using important landmarks up to 1000, including the factors of 1000 and multiples of those factors
- Developing strategies for adding and subtracting numbers in the hundreds
- Estimating quantities up to 1000

Assessment Resources

Teacher Checkpoint: Find the Numbers (p. 39)

Put One in Each Corner and One in the Middle (Dialogue Box, p. 40)

Assessment: Make Your Own Related Problem Set (p. 47)

Estimating Large Quantities (Teacher Note, p. 49)

Playing Close to 100 (Teacher Note, p. 50)

Materials

Construction paper or oak tag
Hole punch or stapler
Paper fasteners or yarn
Chart paper
Overhead projector, transparencies, and pens
Calculators
Container of beans
Measuring materials (spoons, scoops, paper cups)
Balances
Numeral cards
Student Sheets 13–20
Teaching resource sheets

Investigation 4 ▪ Making a 10,000 Chart

Class Sessions	Activities	Pacing
Sessions 1, 2, and 3 (p. 54) 10,000 SQUARES ON THE WALL	100 for Every Student Completing the 10,000 Chart What Can We See on the 10,000 Chart? Choosing Student Work to Save Homework: Close to 1000	minimum 3 hr

Ten-Minute Math ▪ Counting Around the Class

Mathematical Emphasis

- Reading, writing, and sequencing numbers in the thousands
- Getting a sense of the magnitude of 10,000
- Understanding the structure of 10,000 (for example, that it can be constructed of 10 thousands or 100 hundreds)
- Adding and subtracting multiples of 100 to numbers in the thousands

Assessment Resources

Choosing Student Work to Save (p. 58)

Materials

Calculators

Transparency of blank 100 chart

Student Sheet 19

Teaching resource sheet

MATERIALS LIST

Following are the basic materials needed for the activities in this unit. Many of the items can be purchased from the publisher, either individually or in the Teacher Resource Package and the Student Materials Kit for grade 4. Detailed information is available on the *Investigations* order form. To obtain this form, call toll-free 1-800-872-1100 and ask for a Dale Seymour customer service representative.

Snap™ Cubes (interlocking cubes): 100–200 per student pair

Calculators: at least 1 per pair of students

Numeral cards (available in Teacher Resource Package or separately; or use blackline master to make your own)

Chart or poster paper: several sheets per student

Chips (counters or centimeter cubes) that fit on the 100 chart squares: 5 per student

Balances (optional)

Construction paper or oak tag: 4 sheets per student (optional)

Beans or other small objects in containers

Hole punch, paper fasteners or yarn for putting books together (optional)

Envelopes: 1 per student

Measuring materials: a collection of spoons, scoops, and paper cups with which students can measure quantities of small objects

Overhead projector, transparencies, and pens

A cleared space in your room for the 10,000 display: wall space 8 feet wide by 10 feet high to hang the 100 charts in a 10-by-10 array

Crayons or markers

The following materials are provided at the end of this unit as blackline masters. A Student Activity Booklet containing all student sheets and teaching resources needed for individual work is available.

Family Letter (p. 66)

Student Sheets 1–20 (p. 67)

Teaching Resources:

- 300 Chart (p. 82)
- One-Centimeter Graph Paper (p. 83)
- 101 to 200 Bingo Board (p. 84)
- How to Play 101 to 200 Bingo (p. 85)
- Numeral Cards (p. 86)
- Tens Cards (p. 89)
- How to Play Close to 100 (p. 100)
- How to Play Close to 1000 (p. 101)
- 100 Chart (p. 102)
- Blank 100 Chart (p. 103)

Practice Pages (p. 105)

Related Children's Literature

Mathis, Sharon Bell. *The Hundred Penny Box.* New York: The Viking Press, 1975.

Pittman, Helena Clare. *A Grain of Rice.* New York: Hastings House, 1986.

Seuss, Dr. *The 500 Hats of Bartholomew Cubbins.* New York: The Vanguard Press, 1938.

ABOUT THE MATHEMATICS IN THIS UNIT

An important part of students' mathematical work in the elementary grades is building an understanding of the base ten number system. This unit provides activities that develop knowledge about important *landmarks* in that system—numbers that are familiar landing places, that make for simple calculations, and to which other numbers can be related.

Because our number system is based on powers of ten, the numbers 100 and 1000 and their multiples are especially important landmarks. When solving real problems, people with well-developed number sense draw on their knowledge of these important landmarks. For example, think about how you would solve this problem before you continue reading:

> If there are about 25 students in each class and 17 classes in our school, about how many students are there altogether?

Many people would use their knowledge that there are four 25's in every 100 to help them solve this problem mentally. Rather than multiplying 17 by 25, they will think something like this: "There are four 25's in 100; there are eight in 200, 12 in 300, 16 in 400—and one more 25 makes 425."

Knowledge about 100 and 1000 and their multiples and factors is part of the basis of good number sense. As students learn about 100 and 1000, how to take these numbers apart into their factors, and how to use them to construct other numbers, they gain the knowledge they need to develop their own strategies to solve problems. They learn what happens when you add or subtract multiples of 10 and 100. They develop good estimation strategies and are less likely to make errors that result from the use of faulty algorithms.

For example, a student who has sound knowledge about these relationships can easily solve a problem such as this:

$$629 + 72$$

A student might add 70 to 629 by counting on by 10's: 629, 639, 649, 659, 669, 679, 689, 699, then add on the remaining 2 to get 701. Or a student might recognize that 629 is close to 630 and that 30 and 70 add to 100, so take one from 72, and use it to make 629 into 630, add 70 to 630 to get 700, and add one more (from the 72) to make 701. Finally, another student might notice that these numbers are related to 25 and 75: 625 and 75 would make 700, subtract the extra 3 added to the 72 to make it 75 to get 697, and add on the extra 4 from the 629 to get 701. None of these methods involves "carrying," which students often use as a rote procedure they don't really understand and frequently misapply. In fact, all these strategies move from left to right, working with the biggest numbers first, an approach that often leads to more accurate calculation.

Sound mental strategies such as those described above are used by people who are fluent in computation. When students use strategies based on their own good number sense, they are more likely to solve problems accurately and to make sure their solutions make sense. They tend to see quantities as whole quantities, not individual digits (for example, 629 is about 30 bigger than 600, *not* 629 is a 6 and a 2 and a 9 in a row), so they are more likely to notice if their results are unreasonable.

It is also important that students develop some sense of the magnitude and relationships of these numbers: How big is 100? 1000? 10,000? How many hundreds are in 10,000? How far is it from 3000 to 7500? If you added 300 to 7800, what would the result be? For this reason, we spend time in this unit estimating large quantities, as well as making a 1000 book and a 10,000 chart, so students can begin to visualize these important relationships.

At the beginning of each investigation, the Mathematical Emphasis section tells you what is most important for students to learn during that investigation. Many of these mathematical understandings and processes are difficult and complex. Students gradually learn more and more about each idea over many years of schooling. Individual students will begin and end the unit with different levels of knowledge and skill, but all will gain greater knowledge about hundreds and thousands and develop strategies for solving problems involving these important numbers.

Throughout the *Investigations* curriculum, there are many opportunities for ongoing daily assessment as you observe, listen to, and interact with students at work. In this unit, you will find three Teacher Checkpoints:

Investigation 1, Session 3
Jumping to 100 (p. 15)

Investigation 2, Sessions 2–4
Boxes for 200 (p. 27)

Investigation 3, Session 1
Find the Numbers (p. 39)

This unit also has two embedded assessment activities:

Investigation 2, Session 5
Problems with Landmarks in the Hundreds (p. 34)

Investigation 3, Sessions 3–5
Make Your Own Related Problem Set (p. 47)

In addition, you can use almost any activity in this unit to assess your students' needs and strengths. Listed below are questions to help you focus your observation in each investigation. You may want to keep track of your observations for each student to help you plan your curriculum and monitor students' growth. Suggestions for documenting student growth can be found in the section About Assessment.

Investigation 1: Working with 100

- How comfortable are students counting by factors of 100? What strategies do students use to find factors of 100? Are students systematic in their approach? Can they find all the factors? Can they explain why a number is or is not a factor of 100?
- How do students connect and recognize pairs of factors? How do they connect factor pairs to the dimensions of rectangles for a given number?
- How do students find the difference between numbers under 100? Do they use landmark numbers? Do they count up (or back) from one number to the other? Do they use combinations they already know?
- How do students make conjectures about factors? Do students reason about what they know to help them find additional solutions? (For example, "25 would work because 25 works for 100, so it has to work for 200. It's just like counting to 100 twice.")

Investigation 2: Exploring Multiples of 100

- How do students explore multiples of 100? Do they use their knowledge of 100? (For example, if there are five 20's in 100, there are ten in 200 and fifteen in 300.) How comfortable are they skip counting by factors of 100? by larger numbers? How do they keep track?
- How do students make sense of standard multiplication and division notation? Do they use their knowledge of factors to interpret and make sense of the problem and what it is asking? Can they use standard division or multiplication notation to express a problem and its solution?
- How do students add and subtract multiples of 10 to numbers in the 100's? What strategies do they use to solve other addition and subtraction problems? Do they use known relationships? landmark numbers? Can students extend the solution of a known problem to solve a more difficult problem?
- How clearly do students communicate their strategies orally? on paper? How comfortable are they in doing so? What combination of words, pictures, and numbers do they use?

Investigation 3: How Much Is 1000?

- How comfortable and accurate are students with reading and writing numbers through 1000? Do they have a sense of the sequence of numbers? How do they locate where a number is to be written in their 1000 book?
- What evidence do you have that students are developing a sense of the magnitude of multiples of 100 up to 1000?
- How easily do students identify and use landmark numbers up to 1000? Do they use factors of 1000? multiples of those factors? How do students make connections between multiples of 100 and multiples of 1000?
- What strategies are students developing to add and subtract numbers in the hundreds? Do they use landmarks? Do they count up (or back) from a number? Do they use familiar combinations they already know?

- How do students estimate quantities of objects in the hundreds? How do they refine their predictions? How do they compare predictions with actual outcomes?

Investigation 4: Making a 10,000 Chart

- How comfortable and accurate are students with reading and writing numbers in the 1000's? Do they have a sense of the sequence of numbers? How do they locate where a number is to be written in their 1000 book?
- What evidence do you have that students are developing a sense of the magnitude of 10,000? How do they locate numbers on the 10,000 chart?
- How do students make sense of and understand the structure of 10,000? (For example, 10,000 can be constructed of 10 thousands, 100 hundreds, or 1000 tens.)
- How do students add and subtract multiples of 100 to numbers in the thousands? What strategies do they use? Do they use landmarks? skip counting? familiar combinations?

In the *Investigations* curriculum, mathematical vocabulary is introduced naturally during the activities. We don't ask students to learn definitions of new terms; rather, they come to understand such words as *factor*, *area*, and *symmetry* by hearing them used frequently in discussion as they investigate new concepts. This approach is compatible with current theories of second-language acquisition, which emphasize the use of new vocabulary in meaningful contexts while students are actively involved with objects, pictures, and physical movement.

Listed below are some key words used in this unit that will not be new to most English speakers at this age level but may be unfamiliar to students with limited English proficiency. You will want to spend additional time working on these words with your students who are learning English. If your students are working with a second-language teacher, you might enlist your colleague's aid in familiarizing students with these words before and during this unit. In the classroom, look for opportunities for students to hear and use these words. Activities you can use to present the words are given in the appendix, Vocabulary Support for Second-Language Learners (p. 63).

layer, box Students explore factors of 100 and 200 by finding different ways to pack cubes (or another small square object) in a box. After working on single-layer configurations, they investigate arrangements with more than one layer.

jump, ahead, land exactly Students explore multiplication by participating in jumping races. Students also figure how many jumps it would take to land on a piece of paper placed on the floor.

Multicultural Extensions for All Students

Whenever possible, encourage students to share words, objects, customs, or any aspects of daily life from their own cultures and backgrounds that are relevant to the activities in this unit. For example:

- Students who have coins from their countries of origin can bring them to show to the class. They might make a poster showing equivalencies of the coins in these monetary systems.
- When students are making up their own word problems, encourage them to write problems that are based on aspects of their cultures—foods, games, and sports that involve teams, and so forth.

Investigations

INVESTIGATION 1

Working with 100

What Happens

Session 1: Ways to Count to 100 Students skip count by 2's and by 6's on the 100 chart and discuss patterns they see in the multiples of 2 and 6. Then they find factors of 100, skip count by those factors on miniature 100 charts, and develop conjectures about how to tell which numbers are and aren't factors of 100.

Session 2: 100 in a Box Students make rectangles of different shapes using 100 cubes. They make these "boxes" first with only one layer of cubes (for example, a 10 by 10 array), then with more than one layer (for example, four layers of 5 by 5 squares).

Session 3: Moving Around on the 100 Chart Students are introduced to the idea of "landmarks" in the number system. They explore the difference between various two-digit numbers and 100, one of these important landmarks.

Mathematical Emphasis

- Finding and counting by factors of 100
- Recognizing factor pairs (for example, 4 rows of 25 cubes make 100 and 25 rows of 4 cubes make 100)
- Using landmarks to find differences between numbers under 100 (for example, the difference between 48 and 100 is 52 because from 48 to 50 is 2 and then it's 50 more to 100)
- Making conjectures about factors of 100

What to Plan Ahead of Time

Materials

- Overhead projector, transparency pen (Session 1)
- Chart paper: 3 large sheets (Session 1)
- Interlocking cubes: at least 100 per pair (Session 2)
- Chips (counters or centimeter cubes) that fit on 100 chart squares: 3 per student (Session 3)

Other Preparation

- Duplicate student sheets and teaching resources (located at the end of this unit) in the following quantities. If you have Student Activity Booklets, copy only the items marked with an asterisk.

  For Session 1

  Student Sheet 1, Miniature 100 Charts (p. 67): 3–5 per pair

  Student Sheet 2, Factors of 100 (p. 68): 1 per pair

  Family Letter* (p. 66): 1 per student. Remember to sign it before copying.

  Student Sheet 3, More on Factors of 100 (p. 69): 1 per student (homework)

  100 chart (p. 102): 2 per student (class), 1 per student (homework), plus extras*

  For Session 3

  Student Sheet 4, How Far to 100? (p. 70): 1 per student

  Student Sheet 5, Jumping on the 100 Chart (p. 71): 1 per student (homework)

  100 chart (p. 102): 1 per student (class), 1 per student (homework)

- Students will use the 100 chart as a standard material throughout the unit. Make sure there is always a supply of these readily available for all students.
- Make three overhead transparencies of the 100 chart (p. 102). (Session 1)
- If you plan to provide folders in which students will save their work for the entire unit, prepare these for distribution during Session 1.

Session 1

Ways to Count to 100

Materials

- 100 chart (2 per student, class; 1 per student, homework; plus extras)
- Transparency of 100 chart (3 copies)
- Student Sheet 1 (3–5 per pair)
- Student Sheet 2 (1 per pair)
- Student Sheet 3 (1 per student, homework)
- Family letter (1 per student)
- Chart paper (3 large sheets)
- Overhead projector and transparency pen

What Happens

Students skip count by 2's and by 6's on the 100 chart and discuss patterns they see in the multiples of 2 and 6. Then they find factors of 100, skip count by those factors on miniature 100 charts, and develop conjectures about how to tell which numbers are and aren't factors of 100. Students' work focuses on:

- finding and counting by factors of 100

Activity

Skip Counting on the 100 Chart by 2's and 6's

Note: If you are doing the full-year *Investigations* curriculum for grade 4 and your class has already done the unit *Arrays and Shares*, your students may already be very comfortable with skip counting on the 100 chart. If so, this activity can be done as a quick review of skip counting. If your students are not familiar with skip counting, you may need to spend more time with this activity; have students skip count by and highlight multiples of numbers other than 2 and 6 on the 100 chart.

Using the overhead projector and a transparency of the 100 chart, ask students to help you count by 2's. As students tell you the next number in the series (2, 4, 6, 8, 10, 12, 14, . . .), mark it by circling it or putting an X on it. Count up to about 30. Ask:

How did you know where to start? How do you know what comes next?

Encourage a variety of responses. Even though one student has already given a correct answer, other students can tell their own ways of thinking about it. Some students will notice patterns on the 100 chart ("It's only the even numbers," "You just keep going under the ones you've already done," "You skip every other number"). See the **Dialogue Box**, Counting by 2's and 6's (p. 9), for some observations fourth graders have made.

Activity

Highlighting 100 Charts

Highlighting 100 Charts for 2's and 6's Students now complete the 2's pattern on their own 100 chart by circling or marking in some way the multiples of 2. When students finish the 2's, they count by 6's on another 100 chart. As you circulate, make sure students know how to start the 6's. Require students to compare charts with one another and, if there are discrepancies, to try to figure out why there are differences. This step is critical. Some students will be misled by false patterns or may simply miscount. Double-checking by comparing and discussing their work with someone else is an important part of doing mathematics. Have plenty of copies of the 100 chart on hand so students feel free to do another one if they "mess up."

When students have completed both the 2's and the 6's, they work with one or two other students to compare the 2's chart to the 6's chart. Ask them to write down three things that are either similar or different about the 2's and the 6's.

❖ **Tip for the Linguistically Diverse Classroom** Have limited-English-proficient students show their thoughts. For example:

Students can point to numbers that appear on both charts.

Students can point to a number on one chart and then show how it does not appear on the other chart.

Students can point to 100 on the 2's chart to show that the 2 lands exactly on 100; then to 100 on the 6's chart to show that the 6 does not.

Discussion: Comparing the 2's and 6's Charts When students are finished with both charts, bring the class back together. Using another numbered 100 chart transparency, ask students to help you skip count by 6's. Mark the numbers up to 60.

Where did you start? What are some ways you can tell what number comes next?

Responses might include "Count 6 more from 60," "It's the next one in the diagonal, 42, 54, 66," "The numbers always go in a pattern—6, 2, 8, 4, 0, 6, 2, 8, 4, 0 [*this student is referring to the pattern of the units digit*]—so you know the next one after 60 is going to end in 6" (see the **Dialogue Box**, Counting by 2's and 6's, p. 9).

With the students, finish the 6's pattern on the transparency. When you get to 84, again ask them how they can predict the next number. When the counting is completed, ask students what they can see in the 6's pattern. Continue the discussion with these questions:

What are some things that are the same about the 2's pattern and the 6's pattern? What are some differences?

Students can report what they wrote down in their groups.

Activity

What Are All the Factors of 100?

Give each pair of students two copies of Student Sheet 1, Miniature 100 Charts, and one copy of Student Sheet 2, Factors of 100. Start this activity by saying something like this:

One difference between the 2's and the 6's is that when we skip count by 2, we land exactly on 100; but when we use 6, we do not. You can divide the 100 chart into 2's with nothing left over, but you can't do that with 6's.

Work with your partner to find other numbers you can count by and land exactly on 100. Use the miniature 100 charts to check each number. If you land on 100, write how many jumps it took. On Student Sheet 2, make a list of all the numbers you try, and mark whether or not your counting landed exactly on 100.

See if you can find all the possible numbers you can count by and land exactly on 100.

As you circulate, ask students to show or tell you how they know a particular number works—that is, counting by that number lands exactly on 100. Also, make sure they are recording on Student Sheet 2 the numbers that don't work. Encourage them to keep checking if they have not found all the possible numbers. Some students may not realize they can count by numbers bigger than 10, so remind them they can do so. Give students additional copies of Student Sheet 1 as needed. If students find all the factors of 100 (1, 2, 4, 5, 10, 20, 25, 50, 100) without trying any others on the miniature charts, they will need at least three copies of Student Sheet 1 to show them all.

Activity

Discussion: Do We Have All the Factors of 100?

With the class, make a list on chart paper of all the factors of 100 they have found. Don't comment on whether they have found them all. Briefly discuss questions such as these:

What numbers did you try that didn't work? How do you know 3 doesn't work? How did you find out that 20 worked? Were there any numbers that surprised you? Did anyone try 49? Why not? Could it work? Why or why not?

This discussion provides a good opportunity to use the word *factor*. Students who have completed the unit *Arrays and Shares* will already be familiar with this term. Explain that a factor of 100 is any number you can count by to get to 100, including 1 and 100. Make sure they understand that all numbers have factors, not just 100. For example, the factors of 12 are 1, 2, 3, 4, 6, and 12. You don't need to insist that students use the word *factor*. Rather, use the word yourself naturally, in context, so students will gradually learn its meaning and feel comfortable using it themselves. See the **Teacher Note**, Introducing Mathematical Vocabulary (p. 10).

Then focus on these questions:

Do we have all the factors of 100? How can you tell if you have found them all or not? Do you have to try every number between 1 and 100? Are there any numbers you're sure you *don't* have to try?

As students respond, challenge the group about how they know that particular numbers are or are not factors of 100:

How do you know 32 doesn't work? Who can prove it? How can you know that nothing in the 30's works?

On an overhead transparency of the 100 chart, circle the factors students have found and cross out all the numbers everyone agrees can be eliminated as factors of 100. At the same time, list on chart paper students' conjectures about how to tell which numbers are *not* factors of 100. Try to move the conversation from a focus on *individual* numbers that have been tried ("I know 26 doesn't work because I tried it") to more general notions about how to eliminate numbers that are not factors ("You don't have to test anything above 90 because you can't multiply 90 by any whole number except 1 and get a number less than 100," "You don't have to test anything with a 9 at the end because if you multiplied it, you wouldn't get a zero—except 10 × 9—so you can't get 100"). Whether students' conjectures are true or not, include them in the list. Throughout the unit, students can continue trying to prove or disprove them. Making conjectures and proving them is a difficult and unfamiliar task for students, so allow plenty of time for this discussion.

At the end of the discussion, list the circled factors from the overhead transparency on chart paper. Then return to this question:

Do we have all the factors of 100? How sure are you that we have them all?

Keep the list of factors and conjectures posted for future reference. If the list of factors is not yet complete or if your students are not sure whether they have all the factors, tell them you will keep the list posted. Students can propose other factors of 100 to add to it later in the unit. You may also want to choose and cut out one of the completed 100 charts for each factor to post near your list.

Session 1 Follow-Up

Homework

More on Factors of 100 For homework, send home Student Sheet 3, More on Factors of 100, and a 100 chart. Students answer questions about possible factors of 100 and explain their reasoning. Send home the family letter or *Investigations* at Home with this sheet.

Extensions

A Book of 100 Charts Each student makes a book of 100 charts. Each 100 chart shows counting by one of the factors of 100. The class can make miniature books with one page for each factor; for each factor they can color in the pattern of that factor on a 10 by 10 grid of graph paper with very small squares (1/4 inch or 0.5 centimeter). They can cut out each grid and paste it in construction paper books. These provide an excellent visual reference for the factors of 100.

Finding Factors Students find factors of other numbers. Interesting ones to work on are 24, 30, 36, 40, 48, 60, and 90.

Counting by 2's and 6's

This discussion took place while students were doing the activity Skip Counting on the 100 Chart by 2's and 6's.

How do you know what comes next?

Emilio: By skipping a number.

Teresa: I was doing it by 2's, but you could just count *down*—like 2, 12, 22, 32.

Are there any other ways that people are thinking about it?

Marci: You keep doing it over, I mean the 2's—22, 24, 26. It's always 2, 4, 6, 8, 0 on every line.

Kyle: Adding.

What do you mean by that?

Kyle: Well, like 2 plus 2 is equal to 4, 4 plus 2 is equal to 6, and on like that.

[*Later*]

Does anyone notice anything about the patterns you've done so far?

Joey: [*looking at the 2's*] There are 50 even numbers.

Rafael: It goes straight down, not diagonal.

On the 2's?

Karen: On the 6's they all go diagonally.

What do you mean by diagonally?

Karen: You skip one, then go to the next one [*demonstrating on the overhead what she means by diagonal, by showing how to move from 42 to 54*].

Irena: It's like the way you move the knight in chess.

Any more patterns you see on the 6's?

Jesse: [*pointing to each row of numbers*] There's one here, then two here, then two, then one again, then two, two.

Kim: You color in a 10, then skip two 10's.

1	**2**	3	**4**	5	**6**	7	**8**	9	**10**
11	**12**	13	**14**	15	**16**	17	**18**	19	**20**
21	**22**	23	**24**	25	**26**	27	**28**	29	**30**
31	**32**	33	**34**	35	**36**	37	**38**	39	**40**
41	**42**	43	**44**	45	**46**	47	**48**	49	**50**
51	**52**	53	**54**	55	**56**	57	**58**	59	**60**
61	**62**	63	**64**	65	**66**	67	**68**	69	**70**
71	**72**	73	**74**	75	**76**	77	**78**	79	**80**
81	**82**	83	**84**	85	**86**	87	**88**	89	**90**
91	**92**	93	**94**	95	**96**	97	**98**	99	**100**

1	2	3	4	5	**6**	7	8	9	10
11	**12**	13	14	15	16	17	**18**	19	20
21	22	23	**24**	25	26	27	28	29	**30**
31	32	33	34	35	**36**	37	38	39	40
41	**42**	43	44	45	46	47	**48**	49	50
51	52	53	**54**	55	56	57	58	59	**60**
61	62	63	64	65	**66**	67	68	69	70
71	**72**	73	74	75	76	77	**78**	79	80
81	82	83	**84**	85	86	87	88	89	**90**
91	92	93	94	95	**96**	97	98	99	100

Teacher Note — Introducing Mathematical Vocabulary

This unit provides the opportunity for introducing several important mathematical words naturally. Introduce these words by beginning to use them yourself and explaining what you mean by them, but don't insist that students use them. If the introduction of a vocabulary word is preceded by activities that make its definition clear, students enjoy knowing an "adult" word to refer to a new concept they have learned.

Multiple *Multiple* is a natural word to introduce to students as they count by 2's (or any other number) and circle those numbers on the 100 chart. During the activity, refer to the numbers you have circled as the "multiples of 2." You might ask students if they have other names for these numbers or for this way of counting, such as the "2's table" or "counting by 2's."

Factor *A factor* of a number is a number that can be divided evenly into that number. For example, 2 is a factor of 4, 6, 8, and all the even numbers; 1, 2, 4, 8, and 16 are all factors of 16. The idea of factors comes up naturally when students make and use arrays. The dimensions of each array are factors of the total number in the array.

Even and **Odd** These words will come up in students' descriptions of patterns on the 100 chart. Don't assume students know exactly what they mean by these words. Some students believe that an even number has only even factors: "No, 3 isn't a factor of 24 because 3 isn't even." This is a good conjecture to have students investigate.

Row and **Column** When talking about their ways of working with the 100 chart, students often confuse the words *row* and *column*, describing a pattern as going "down the row" rather than "down the column." This may be a good opportunity to talk about the difference, since using *row* for both (as students often do) makes communication more difficult. However, do not insist students use these words in the conventional way as long as they can explain or demonstrate what they mean. Remembering the difference can be honestly confusing, and focusing on getting the words right may obscure the good mathematical thinking a student is doing. Rather, keep using them yourself so students continually hear them used correctly in context.

columns ↓

rows →

1	2	3	4	5	6	7	8	9	10
11	12	13	14	15	16	17	18	19	20
21	22	23	24	25	26	27	28	29	30
31	32	33	34	35	36	37	38	39	40
41	42	43	44	45	46	47	48	49	50
51	52	53	54	55	56	57	58	59	60
61	62	63	64	65	66	67	68	69	70
71	72	73	74	75	76	77	78	79	80
81	82	83	84	85	86	87	88	89	90
91	92	93	94	95	96	97	98	99	100

100 in a Box

What Happens

Students make rectangles of different shapes using 100 cubes. They make these "boxes" first with only one layer of cubes (for example, a 10 by 10 array), then with more than one layer (for example, four layers of 5 by 5 squares). Students' work focuses on:

- using factors of 100 to build rectangular solids with a total of 100 cubes
- keeping track of and recording their findings

Materials

- Interlocking cubes (at least 100 per pair)

Activity

Building with 100 Cubes

Introduce this activity by describing a scenario such as the following:

A company that makes caramels (or brownies or erasers or anything else that is roughly square) **is trying to decide what size box they should use. They want to have 100 caramels in a box and want to have only one layer in each box. Before they decide, they want to figure out what all the possible rectangular arrangements are for exactly 100 caramels with none left over and no extra space. You will be using 100 interlocking cubes to show the company different arrangements for their caramels.**

❖ **Tip for the Linguistically Diverse Classroom** Show a real example of whatever item you decide to have students "box" for this activity.

You may want to draw on the board or overhead a few rectangles of different shapes (without indicating the exact dimensions) to give your students the idea of what you mean:

You might make a long, skinny rectangle like this [*sketch one*] **or a fatter rectangle like this** [*sketch a different one*]**, but remember that you have to use exactly 100 cubes for each arrangement.**

If students are still not sure what the task is, show them a 10 by 10 rectangle made with 10 rows of interlocking cubes with 10 cubes in each row.

Students work in pairs with interlocking cubes. They record (drawing or writing) all the ways they can find to make a rectangle with the 100 cubes. They need to record as they make a rectangle since they will have to take apart each array to build another. We have deliberately not provided a student sheet for this activity so students have the opportunity to think about how to keep track and record in their own ways. Several important mathematical ideas are likely to come up as students build. Be alert for students who are thinking about these ideas, and ask questions to push their thinking further. See a description of these ideas and sample questions in the **Teacher Note**, Boxes for 100 (p. 14). You may also find that some of your students have a much more difficult time with this activity than with skip counting; see the **Teacher Note**, Students' Difficulties Making Arrays (p. 13).

Making Boxes with More Than One Layer If some student pairs have found all the one-layer boxes they can make, suggest they see if there are any boxes with more than one layer. Once a few students have made some boxes with more than one layer, you might stop and introduce this idea to the whole class (or just continue suggesting it individually to student pairs as they are ready):

Emilio and Katie made a box like this one. It has 100 cubes, but it has more than one layer. Well, the caramel company had been thinking only about flat one-layer boxes, but when they saw Emilio and Katie's idea, they decided they wanted to think about other kinds of boxes, with more than one layer. So if you think you have found all the flat boxes, see if you can make some multilayer boxes like this one with exactly 100 cubes.

For some students, just finding one of these may be a good challenge. Others may enjoy being challenged to find all the possible multilayer boxes. Only boxes that are more than one layer high, wide, and deep count as multilayer. Depending on what is defined as a "different" box—see the **Teacher Note**, Boxes for 100 (p. 14)—there are either three different multilayer boxes (if orientation doesn't matter) or seven different ones (which rectangle is on the bottom does matter).

As students make boxes with more than one layer, they need to continue to keep track of their work. If students don't know how to write down what they have made, ask them to describe it to you in words. As they describe it out loud, they will clarify their own thinking about what they have made; then they can use some of the words they said aloud to record their work (for example, "We have 5 rectangles on top of one another; each one has 4 rows of 5" could be written as "5 rectangles with 4 rows of 5"). They may also check their constructions on the calculator: multiply 4 by 5 to get one layer; then multiply the result by 5 (the number of layers). We have found that fourth graders do not readily understand how the notation $4 \times 5 \times 5$ describes a three-dimensional construction. However, if this idea comes *from the students,* you can certainly let them know that this is one way mathematicians describe such objects.

Displaying Students' Solutions Make a display on a table, a desk, or the floor showing one of each kind of box made of 100 cubes. (You won't need all the cubes again for a couple of sessions.) If students find new ones over the next couple of days, they can add to this display. One group of students can be responsible for making this display; they can be authorized to take a differently shaped box from each group of students to put in the display and to make a label giving its dimensions.

Students' Difficulties Making Arrays

Teacher Note

Making arrays for 100 and other numbers has proven quite difficult for some fourth graders. Even some students who easily found factors of 100 through skip counting may approach making "boxes" (rectangular arrays) as a completely new problem. Expect a real range within your classroom—from students who have great difficulty finding even one array to students who are quickly making three-dimensional arrays. Expect more experimentation and less certainty than occurred during the skip counting activities. Here is a sample of a conversation between two students working on this activity:

Tyrone: I'll do 40 by 3.

Lesley Ann: No, that's 120.

Tyrone: How about 30 by 3?

Lesley Ann: That's 90.

Tyrone: Forty by 3 . . . 40 by 40? 800? Let me try it: 10, 20, 30, 40 [*counting cubes*].

Lesley Ann: What are you trying to do?

Tyrone: Forty by 40 [*laying out two rows of 40*].

Lesley Ann: What is 40 by 40?

Tyrone: That's 160.

Tyrone seems confused about dimensions and their relation to a whole. He may not have any visual image of what is meant by a phrase such as "40 by 40" and may be thinking of it as "40 and 40." Where you find natural places to talk about this, introduce language such as "50 by 2," "50 rows of 2," and the notation 50×2 in contexts where students can see what these phrases mean: "I see you tried a 40 by 2 rectangle; did that one have 100 cubes? Did the one with rows of 25 work? Oh, I see, it came out to be 25 by 4."

We have also seen students who attempt first to outline a rectangle with cubes, then to fill it in. They focus on a frame with empty space inside it. This method takes the focus off the view of the rectangle as a certain number of rows of a certain number of cubes. Ask these students to fill in their rectangle, then talk with them about the number of rows they have, the number in each row, and the total number of cubes they have used.

Teacher Note — Boxes for 100

While making their boxes of 100, students may bring up some important mathematical ideas:

- The dimensions of their boxes, both two-dimensional and three-dimensional, are the same numbers they found when they skip counted on the 100 chart with numbers that land exactly on 100—the *factors* of 100. Here are questions to pose: "Why do you think the same numbers appear in both situations? Here's a box of 2 rows of 50. What does 2 rows of 50 have to do with what you can do on the 100 chart? On the 100 chart, you found out you could skip count by 4's and land on 100, so 4 is a factor of 100. Is there anything the same about skip counting by 4 and any of the boxes you made?"
- Factors come in pairs. If you find that 4 is a factor of 100, then there is another number which, when multiplied by 4, gives you 100 (in this case, 25). This number is also a factor of 100. Here are questions to ask: "Is this always true? Do factors always come in pairs? Why do they? If 4×25 is equal to 100, does 25×4 have to be equal to 100? Is this true for other numbers? What about 10? Is there another factor of 100 it is paired with?"
- Five rows of 20 and 20 rows of 5 make the same-shape box. Here is a question to pose: "Should we consider these boxes to be different or the same?" There is not one answer to this question: Both sides of the question can be defended. For example, students might contend that because you can simply rotate the box and make it look exactly like the other box, they are the same—it wouldn't make any difference to the caramel company which way people held the box. On the other hand, someone might say that if you spread the rows apart, 5 rows of 20 (a few very long rows) looks very different from 20 rows of 5 (a lot of short rows). If the caramel company puts dividers between the rows, five 20's and twenty 5's might look different to the eye.

The same issue will come up when students make multilayered boxes: "Is a box made of 2 layers that are each 5 by 10 the same as a box made of 5 layers that are each 10 by 2?" Again, there are arguments for both points of view, although the argument for considering them as different, at least in this context of boxes for caramels, may be even stronger than for one-layer boxes. After all, a rectangular solid with a 5 by 10 rectangle as its base really does look different from one with a 2 by 10 rectangle as its base.

If boxes with the same dimensions are considered the same no matter which part is on the bottom, there are three different multilayer boxes: 2 by 2 by 25, 2 by 5 by 10, and 4 by 5 by 5. If unique boxes are determined by the rectangle on the bottom, there are seven different boxes with the bottoms: 2 by 2, 2 by 25, 2 by 5, 2 by 10, 5 by 10, 4 by 5, and 5 by 5.

Moving Around on the 100 Chart

What Happens

Students are introduced to the idea of "landmarks" in the number system. They explore the difference between various two-digit numbers and 100, one of these important landmarks. Students' work focuses on:

- finding the difference between any two-digit number and 100
- moving around on the 100 chart by using landmarks (for example, "To get from 48 to 100, I jumped 2 squares to 50, then I knew it was 50 more to 100").
- adding 10 to or subtracting 10 from two-digit numbers (for example, "I jumped by 10's from 35—45, 55, 65, 75, . . .")

Materials

- 100 chart (1 per student, class; 1 per student, homework)
- Chips (counters or centimeter cubes) for 100 chart (3 per student)
- Student Sheet 4 (1 per student)
- Student Sheet 5 (1 per student, homework)

Activity

Teacher Checkpoint

Jumping to 100

Introducing "Landmark" Numbers Briefly discuss the idea of landmarks and students' experience with landmarks in their everyday lives:

What is a landmark? Why are they useful? Have you ever used landmarks or seen anyone in your family use landmarks? Can a tree be a landmark? How about a building or a sign for a business?

Explain that there are important numbers we can use as landmarks to help us tell where we are when we are counting or calculating with numbers.

When you count or use numbers, are there some that seem particularly important to you? If so, which ones? Why are these important?

Students might suggest 2, 5, 10, 50, 100, or others. If they think about money, they might think about 25 as an important landmark. In measuring length, 12 is an important landmark; and in telling how much time has passed, 60 is a landmark.

In our number system, 10, 100, and 1000 are very important numbers. So are numbers that are multiples of these numbers, such as 20, 30, and 40 or 500, 600, and 700. That's why we're going to be spending some time finding out everything we can about numbers like 10, 100, 1000, and 10,000.

Froggy Jumps Each student needs a copy of the 100 chart and three chips (counters or centimeter cubes) small enough to fit inside a box on the chart.

Put one of your chips on 50. We're going to pretend that your chip is a frog named Frogurt. Frogurt always jumps one square at a time, in order of the numbers. How many jumps will Frogurt have to take to get to 100? How do you know? Can you prove it? Can you prove it without counting by 1's?

Students may use the chips in different ways. Some students may use all three—one to mark the starting number, one to mark 100, and one to jump with. Others may use only one chip; others may visualize without actually jumping. When students feel secure that Frogurt would take 50 jumps to get from 50 to 100, ask students to put their markers on 48:

How many jumps will it take Frogurt to get from 48 to 100?

Try a couple more examples until you are sure students understand the task; then give each student a copy of Student Sheet 4, How Far to 100? Students may work together, but they record their work individually.

Discussion: How Many Jumps? When all students have done at least half of Student Sheet 4, ask them to share some of their solutions. Point out strategies that make use of landmarks on the 100 chart, such as multiples of 10 or 25 (for example, "To get from 35 to 100, I said it's 5 to 40 and then 60 more to 100"), and those that make use of knowledge about what happens when you add or subtract a multiple of 10 ("To get from 35 to 100, I jumped by 10's—45, 55, 65, 75, 85, 95—that was 60, and then 5 more"). For examples of the kinds of strategies a class of fourth graders used, see the **Dialogue Box**, Jumping to 100 (p. 17).

Session 3 Follow-Up

Homework

Jumping on the 100 Chart Send home Student Sheet 5, Jumping on the 100 Chart, and a copy of the 100 chart. Students do the problems on Student Sheet 5 for homework. You can modify the assignment appropriately for students by having them do either one or two problems.

❖ **Tip for the Linguistically Diverse Classroom** Read the problems on Student Sheet 5 aloud, making use of pointing and actions whenever necessary. Limited-English-proficient students can prove their answers by either drawing pictures or writing their explanations in their native language.

Jumping to 100

The class discusses its strategies and solutions to the first two problems on Student Sheet 4, How Far to 100?

Pinsuba: I counted starting at the bottom [*on 100*] and went up 10, 20, 30, 40, 50, 60 [*he's now on the square labeled 40*], then counted back 5 [*to the square marked 35*].

Nhat: I started out at 35 and counted down 45, 55, 65, up to 95. Then I just counted over 5.

Sarah: I didn't even use the number chart. I said, "35 plus what equals 100."

How did you know what it was going to be?

Sarah: I know my pluses.

Tyrone: I started at 35 and counted by 2 and went up to 100, and I got here [*pointing to the 99*]. It was 64, and I added 1 and got 65.

B. J.: We did the opposite of Sarah. We did 100 minus 35.

Rikki: Yeah, we did 90, 80, 70—that's 30—and then 5 more—that's 65.

What about the second one? Start at 25. How many jumps would it be to get to 100?

David: It would take 75 jumps.

How did you get that?

David: I said 100, right? I made the 100 like one dollar and made the 25 like 25 cents, and then I knew it would be 75.

Luisa: I had a way of checking my answer. I just took my answer and added it to what I started with, and if it didn't add up to 100, then I'd reject that answer.

How did you get 75?

Luisa: I just counted.

Vanessa: You can just count by 25's and get 100. 25, 50, 75 and you get 100.

Nick: I started at 25 and went diagonally by 11's—11, 22, 33, 44, 55 [*pointing to the numbers 36, 47, 58, 69, 80*]. Then I just go down by 10's—65, 75 [*pointing to 90, 100*].

How did you know if you went diagonally it would be 11?

Nick: I just counted the squares.

I started out at 35 and counted 45,55,65 up to 95. Then I just counted over 5.
I counted starting at 100 and went back by 10's 'til I got to 40, then went back 5.

Exploring Multiples of 100

What Happens

Session 1: Factors of 100, 200, and 300 Students count by 2's or 5's on the 300 chart. They explore how many 4's, 20's, and 25's are in 100, 200, and 300 and write about their strategy for finding the number of 4's (or 20's or 25's) in 300. The students discuss how many 20's are in multiples of 100 up to 1000 and consider how to write a problem such as "How many 20's are in 700?" using standard division notation.

Sessions 2, 3, and 4: Using Landmarks to Add and Subtract Students develop strategies for solving sets of related addition and subtraction problems, using a result from one problem (for example, 100 – 20) to solve an unknown one (for example, 100 – 22). During Choice Time, students choose from several activities: finding factors of 300, making boxes for 200 cubes, doing related problem sets involving numbers to 300, and playing a game of Bingo on the 101 to 200 chart.

Session 5: Solving Problems in the Hundreds Students work on one or two word problems as an assessment of their work with 100, 200, and 300 and the factors of these numbers. For the remainder of the session, students continue with the choices from Sessions 2 through 4.

Mathematical Emphasis

- Using knowledge about the factors of 100 to explore multiples of 100 (for example, if there are four 25's in 100, then there are eight in 200 and twelve in 300)
- Relating knowledge of factors to division situations and to standard division notation (for example, 700 ÷ 20 means "How many 20's are in 700?" and can be solved by skip counting by 20's to 700 or by reasoning that there are four 20's in each 100, so there must be 7 × 4, or twenty-eight 20's, in 700)
- Adding multiples of 10 to and subtracting multiples of 10 from numbers in the hundreds
- Solving addition and subtraction problems by reasoning from known relationships (for example, if 100 minus 20 is 80, then 100 minus 22 is 78)
- Communicating strategies orally and on paper through use of words, pictures, and numbers

What to Plan Ahead of Time

Materials

- Chips (counters or centimeter cubes) for 100 and 300 charts: 3 per student (Sessions 1–4)
- Crayons or markers (Sessions 1–5)
- Interlocking cubes: at least 100 per pair (Sessions 1–5)
- Calculators: 1 per pair (Sessions 2–5)
- Envelopes: 1 per student (Sessions 2–4)

Other Preparation

- Duplicate student sheets and teaching resources (located at the end of this unit) in the following quantities. If you have Student Activity Booklets, no copying is needed.
- Make sure 100 charts, 300 charts, and one-centimeter graph paper are available throughout the unit.

  For Session 1

  Student Sheet 6, Exploring Multiples of 100 (p. 72): 1–3 per student

  300 chart (p. 82): 2 per student

  For Sessions 2–4

  Student Sheet 7, Related Problem Sets (p. 73): 6 of each page, cut into cards

  Student Sheet 8, Factors of 300 (p. 77): 1 per student

  Student Sheet 9, More Related Problem Sets (p. 78): 1 per student (homework)

  Student Sheet 10, Froggy Races 1 (p. 79): 1 per student (homework)

  300 chart (p. 82): 2 per student (homework)

  one-centimeter graph paper (p. 83): 1–2 per student

  101 to 200 Bingo Board (p. 84): 2 per student (class), 2 per student (homework)

  How to Play 101 to 200 Bingo (p. 85): 1 per student (homework, optional)

  Numeral Cards (p. 86): 1 deck per group (class), 1 deck per student (homework)

  Tens Cards (p. 89): 1 deck per group (class), 1 deck per student (homework)

  For Session 5

  Student Sheet 11, Two Problems (p. 80): 1 per student

  Student Sheet 12, Another Problem in Two Ways (p. 81): 1 per student (homework)

- Make card decks for playing 101 to 200 Bingo. The game requires a Numeral Cards deck and a Tens Cards deck. You or students will need to make 4 to 6 decks of each for in-class use. Numeral Cards pages 1, 2, and 3 can be photocopied and cut apart to make a complete Numeral Cards deck. (Or you may use the manufactured Numeral Cards from the grade 4 Investigations kits.) Tens Cards pages 1 and 2 can be photocopied and cut apart to make a complete deck of Tens Cards. Students take home their own copies of these sheets to make a deck to use at home (or give students time to cut out their decks in school and provide an envelope to store the cards). (Sessions 2–4)

Session 1

Factors of 100, 200, and 300

Materials

- Interlocking cubes, 100 charts, and chips remain available
- Student Sheet 6 (1–3 per student)
- 300 chart (2 per student)
- Crayons or markers

What Happens

Students count by 2's or 5's on the 300 chart. They explore how many 4's, 20's, and 25's are in 100, 200, and 300 and write about their strategy for finding the number of 4's (or 20's or 25's) in 300. Students discuss how many 20's are in multiples of 100 up to 1000 and consider how to write a problem such as "How many 20's are in 700?" using standard division notation. Students' work focuses on:

- using knowledge about the factors of 100 to find factor pairs for multiples of 100
- counting by factors of 100 to larger multiples of 100
- interpreting standard division notations as a relationship between a factor and its multiple (for example, $900 \div 20$ can be read as "How many 20's does it take to count to 900?")
- looking for patterns in factors of 300

Ten-Minute Math: What Is Likely? Two or three times during the rest of this investigation, do What Is Likely? in any spare ten minutes you have outside of math class—at the beginning of the day or just before lunch, for example.

Fill a container with two colors of cubes, beads, or beans. At first, use much more of one color than the other. Mix well.

Ask students to predict which color they will get more of if they draw ten objects out of the container. Record their predictions.

Draw ten objects from the container, replacing after each draw. Record the results with a table and tallies.

Compare the results with students' predictions. Discuss what happened, focusing on likely and unlikely results. Then draw another ten objects and compare the two results.

For variations on this activity, see p. 59.

Activity

Skip Counting on the 300 Chart

Give a 300 chart and several crayons or markers to each student, and have a discussion centered on the following questions:

If you skip counted on this chart by 2's, would you land exactly on 200? on 300? How do you know? How many 2's would it take to get to 200? to 300? How do you know?

How about by 5's? How many jumps would it take to get to 200? to 300? How about by 50's?

Each student skip counts by either 5's or 2's on the 300 chart, coloring each multiple of 5 (or 2) on the chart. Briefly ask them what patterns they see.

Can you predict how many 5's (or 2's) it would take to get to 400?

Activity

Factors of 200, 300, and Up!

Distribute Student Sheet 6, Exploring Multiples of 100, to each student. Tell students they can choose to use interlocking cubes, 100 charts, or 300 charts for this work. Some students may prefer using three 100 charts together, rather than using the 300 chart, because the squares are bigger. Some students may use only the 100 chart, then predict mentally to the next two hundreds. Although students work in groups, each student should record the group's work on his or her own student sheet. For students who want to work on one or two more factors of 100, distribute an additional Student Sheet 6 for each factor they do. The sentence they write about their work should have enough information to convince someone else how many 4's (or other factor) are in 300. Encourage students to use pictures to help make their arguments clear.

❖ **Tip for the Linguistically Diverse Classroom** Have limited-English-proficient students create sentences that use numbers, pictures, or diagrams. For example:

25 + 25 + 25 + 25 = 100

Four 25's = 100

Some students may want to look at factors of multiples of 100 above 300. If they don't have enough cubes or 100 chart squares to actually count to 400, 500, and so on, they will need to *visualize* and *reason about* what would happen as they count to multiples above 300.

Summing Up Choose one of the factors many students explored. Write the results in a table on the board or the overhead:

Number	How many 20's?
100	5
200	10
300	15

What do you notice about the 20's? How many 20's would there be in 400? How do you know? (Add this result to the table.) **How about higher numbers?** (Add 500 and 600 to the table.) **How many 20's are in 900?** (Pause for all students to think about this.) **Why do you think so?** (Pause until many students have given their own explanations.) **How many 20's are in an "in-between" number such as 640? Is 20 a factor of 640? How do you know?**

Ask students if they noticed patterns for some of the other factors. Look at one more of the patterns together. Extend the pattern beyond 300, and ask some questions about numbers between the multiples of 100—for example, How many 4's are in 540?

As part of this discussion, gradually introduce (or re-introduce) some of the division notation that can be used for problems like these. For example, for the problem "How many 20's are in 900," write on the board these two common notations:

$90 \div 20$ $\qquad\qquad$ $20 \overline{)900}$

These expressions can be read as "How many 20's are in 900?" Continue to use these notations, sometimes using one, sometimes another, to record the problems students encounter during the rest of the unit, so that they are associating the symbols with the meaning of the problems.

Using Landmarks to Add and Subtract

What Happens

Students develop strategies for solving sets of related addition and subtraction problems, using a result from one problem (for example, 100 – 20) to solve an unknown one (for example, 100 – 22). During Choice Time, students choose from several activities: finding factors of 300, making boxes for 200 cubes, doing related problem sets involving numbers to 300, and playing a game of Bingo on the 101 to 200 chart. Their work focuses on:

- adding and subtracting numbers in the 100's
- using landmarks and skip counting to solve problems
- finding factors of 200 and 300
- adding multiples of 10 to and subtracting multiples of 10 from numbers on the 300 chart

Materials

- Interlocking cubes, 100 charts, and 300 charts, and chips for the charts remain available
- Calculators (1 per pair)
- Student Sheet 7 (6 of each page, cut into cards)
- Student Sheet 8 (1 per student)
- Student Sheet 9 (1 per student, homework)
- Student Sheet 10 (1 per student, homework)
- 300 chart (2 per student, homework)
- One-centimeter graph paper (1–2 pieces per student)
- 101 to 200 Bingo Board (2 per student, class; 2 per student, homework)
- How to Play 101 to 200 Bingo (1 per student, homework, optional)
- Numeral Cards (1 deck per group, class; 1 deck per student, homework)
- Tens Cards (1 deck per group, class; 1 deck per student, homework)
- Envelopes (1 per student)
- Pencils, crayons, or markers

Activity

Related Problem Sets

In this activity, you'll introduce related problem sets to the whole class. Students will work on other related problem sets as one choice during Choice Time in this session and again in Sessions 3 and 4. A "related problem set" is a set of problems that helps students use number relationships they know to solve harder problems. If students have done the unit *Arrays and Shares,* they have encountered multiplication problem sets or clusters. In this unit, they will be working on sets of addition and subtraction problems. Solving the problems in the related problem set provides students with ways to think about building a solution for the final problem in the set. For example, in thinking about 100 – 22, you might use the problem 100 – 20 as a way to begin to build a solution. Students can choose any of the problems in the set to help them solve the final problem. In addition and subtraction related problem sets, students often choose one of the problems to provide a first step. Then they figure out how to complete the problem. For example, if 100 – 20 = 80 is a first step to solve 100 – 22, the next step would probably be 80 – 2 = 78. Students are also encouraged to add any problems to the set that they think are useful in solving the problem. Students often think of ways of building solutions that you may not anticipate. Related problem sets are designed to help students develop their number and operation sense and realize that they can apply what they know about number relationships to solve more difficult problems. For more information about related problem sets, see the **Teacher Note,** Related Problem Sets (p. 30), and the **Dialogue Box,** Students' Strategies for Solving Related Problem Sets (p. 32).

Make sure there are interlocking cubes and 100 charts within reach of all students. Write the following set of problems on the board or overhead:

100 – 10 =

100 – 20 =

100 – 2 =

100 – 22 =

Introduce this related problem set by saying something like:

This is a related problem set. The first problems are ones that you might use to help you solve the last problem, 100 – 22. I'd like you to work in pairs to solve the first three problems mentally, with cubes, or with a 100 chart. Later, you'll think about which of these problems you can use as a first step to help you solve the last problem.

Ask student pairs to spend a couple of minutes quietly figuring out the first three problems. They can work on the problems in any order, jotting down their answers. Then, ask students to share their answers to the first three problems in the cluster. Make sure students can justify their solutions by using number relationships they know, by counting, or by using cubes or 100 charts. Then ask:

How could you use any of the first three problems to help you solve 100 – 22? Which problem would you use as a first step? Then what did you do? Who has a different way? Who used a different first step? Did anyone think of another problem you would use to help solve 100 – 22?

At the end of Session 3, bring the class together to share strategies for the related problem sets that you assigned. You may have students who are still trying to do the last problem in the set without thinking about how the other problems give them starting places for building solutions. Use this discussion as a time to encourage students to use good number sense when solving these problems. Compare ways that students solved the final problem by starting with different problems in the set. Also ask students if they thought of any other related problems they could use to start solving the problem.

Present one more problem set and discuss it with the class in the same way. If the first problem set seemed to present the right level of challenge for your students, make up a similar one (such as #1 below). If the first problem set seemed to be easy for your group, choose a more difficult one (such as #2 or #3).

#1	#2	#3
100 – 60 =	100 – 30 =	102 – 10 =
100 – 5 =	100 – 7 =	102 – 4 =
65 + 30 =	37 + 3 =	34 + 60 =
100 – 65 =	**100 – 37 =**	**102 – 34 =**

Remind students:

When you work on related problem sets during Choice Time, find ways to use the first few problems to help you solve the last problem. Try to find at least two different ways to solve the problem. If you think of a different problem that helps you get started solving the last problem, add that problem to the problem set.

❖ **Tip for the Linguistically Diverse Classroom** Limited-English-proficient students can express their methods for solving this problem using numbers, pictures, or diagrams. Have them put a check next to the parts of their method that used other problems in the same set. For example:

✓ 220 + 100 = 320

✓ 220 + 200 = 320 + 100 = 420

228 + 200 = 420 + 8 = 428

Activity

Teacher Checkpoint

Choice Time: Working with 200 and 300

Introducing Choice Time If your students are using the full-year grade 4 *Investigations* curriculum, they are already familiar with Choice Time. You might want to remind them of your particular rules for Choice Time. Spend some time talking with your students about how they make their choices, where the materials are, and where they should work.

Four Choices During the remainder of Session 2 and for most of Sessions 3 and 4, students work largely independently on choice activities. You have just introduced them to related problem sets, which will be one of their choices. The other three are finding factors of 300, which they were introduced to in the previous session; making boxes for 200 (similar to making boxes for 100 in Investigation 1); and playing Bingo on the 101 to 200 chart. These are explained further below. We recommend that you start with the first two choices for the rest of Session 2, add boxes for 200 during Session 3, and then teach 101 to 200 Bingo to small groups during Sessions 3 and 4 (or perhaps to the whole class at the beginning of Session 4). Boxes for 200 is a Teacher Checkpoint, so try to observe all students while they are working on this choice.

We knew that 200 - 135 = 65 because 200 - 30 = 170, so take away 5.

There are more problems and activities than any student will complete. Students work in pairs or groups for all the activities. In some cases, you may decide to have students work individually.

If you set up your choices at stations, show students what they will find at each station. Otherwise, make sure they know where they should get all their materials:

Choice 1: Related Problem Sets—copies of Student Sheet 7 (pp. 1–4, cut into cards); cubes, 100 charts, 300 charts, chips; paper and pencils

Choice 2: Factors of 300—copies of Student Sheet 8, calculators, pencils

Choice 3: Boxes for 200—cubes, 100 charts, 300 charts, chips, calculators, graph paper; paper and pencils

Choice 4: 101 to 200 Bingo—101 to 200 Bingo Board; Numeral Cards; Tens Cards; pencils, crayons, or markers

I did all the first ones by rounding to the highest ten. For example 200-132, 132 is close to 140, and the leftover ones were 8 and the leftover tens were 60, and I put them together and got 68.

Interlocking cubes, 100 charts, 300 charts, and chips should be available for students to work with during Choice Time. Encourage students to use cubes and charts especially for figuring out related problem sets.

Choice 1: Related Problem Sets

Students work in pairs. They select one or more of the problem sets you cut out from Student Sheet 7 (pp. 1–4) and solve the problems in the set, using whatever materials they wish. Decide how you will make these problem sets available to students. You might have a box of them so students can pick up a problem set on which to work, or you might have them posted somewhere in the classroom. Each student writes the problems, the results, and a sentence or two about how he or she solved the last problem in the set on paper or in a notebook. Insist that students take time to record their strategies as well as their solutions. If students are not used to writing about mathematics, ask clarifying questions and tell them to write enough about their strategy so that the reader can understand what they did. Also encourage them to double-check their results by using a second approach or by comparing results with another pair.

Assign one or two problem sets to all students during Session 2; during Session 3, students discuss the problems in these sets as a whole class and share their strategies for solving them (see the next activity, Discussing Related Problem Sets).

Choice 2: Factors of 300

Students find as many factors of 300 as they can and record them on Student Sheet 8, Factors of 300. Although students work in pairs, they record their work individually on the student sheet. Students will find the calculator useful for this activity. As students are working, show them how to use the calculator to skip count and have them teach one another. On most calculators, the = key acts as a constant. Enter a starting number followed by the operation and number you want to use repeatedly.

Then push the = key repeatedly to see how this works. For example, if you

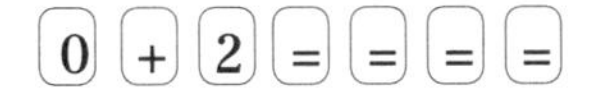

wanted to skip count by 2's, you would press:

Remind students to record on Student Sheet 8 all the numbers they try, not just the ones that work.

Choice 3: Boxes for 200 (Teacher Checkpoint)

Introduce the problem this way:

Remember how we made boxes to hold 100 pieces of caramel? Now the caramel company has increased the number of caramels to 200 in every box. They want to use only one-layer boxes. Your job is to find all the possible one-layer boxes to hold 200 candies. You might also want to recommend which box size you think would be the best.

Students work in pairs or small groups to find boxes for 200. They can use any of the materials that have been introduced, including cubes, 100 charts, 300 charts, and calculators. You can also show students how to use one-centimeter graph paper for drawing their boxes. Their task is to try to find all the arrangements of 200 in one layer and to make a list of the boxes they find. They need to organize their work and make a final list of the box sizes they have found. We do not provide a student sheet for this activity; students are expected to organize and record their own work on paper or in their math notebooks.

Observing the Students

As you watch groups working on this activity, there are several questions you can focus on:

- Are students keeping track of their work carefully?
- Do students double-check their work? (When they think they have 200 cubes, do they actually have 200?)
- Are students becoming fluent in finding factors of a number?
- Are students using tools (cubes, 100 charts, 300 charts, calculators, graph paper) appropriately to help them solve the problem?
- Do students reason about what they know to help them find further solutions (for example, "We knew 25 would work because 25 works for 100, so it has to work for 200; it's just like counting to 100 twice," "At first we didn't find 8, but then we knew that 25 had to go with something, so we figured out how many 25's are in 200, and that was 8. So just to make sure, we built a box with 8's and it worked.")

There are six possible box sizes: 1 by 200, 2 by 100, 4 by 50, 5 by 40, 8 by 25, 10 by 20.

Choice 4: 101 to 200 Bingo

Introduce this game to students in small groups during Sessions 3 and 4 or to the whole class at the beginning of Session 4. If you introduce it to small groups, enlist those who have learned the game to teach other students.

To play 101 to 200 Bingo, each group of students playing together will need a 101 to 200 board, a deck of Numeral Cards, and a deck of Tens Cards.

Before the game starts, each player takes a 1 from the Numeral Cards deck and keeps this card throughout the game. They mix up the deck of Numeral Cards and the deck of Tens Cards and place each in a stack face down on the table.

Now players take turns and help one another with their turns. To determine a play, the player draws two Numeral Cards and one Tens Card. The player uses the 1 card and the other two digits to make a number and adds or subtracts, as indicated, the number on the Tens Card. He or she circles the resulting number on the 101 to 200 Bingo board.

Wild Cards in the Numeral Cards deck can be used to stand for any numeral 0 through 9. Wild Cards in the Tens Cards deck can be used as + or – any multiple of 10 from 10 to 70.

The goal is for the players together to circle five numbers in a row or column (with sides touching) or on a diagonal (with corners touching) to get Bingo.

Activity

Discussing Related Problem Sets

At the end of Session 3, bring the class together to share strategies for the related problem sets you assigned. You may have students who are still trying to do the last problem in the set without thinking about how the other problems give them starting places for building solutions. The problem sets are designed to support students in developing strategies based on using what they know about numbers to solve problems (for example, "Since I know that 100 – 65 = 35, I can reason that 100 – 66 will be 1 less, or 34"). Use this discussion as a time to encourage students to use good number sense when solving these problems. Compare ways students solved the final problem in a set by starting with different problems in the set. Also ask students if they thought of any other related problems they could use to start solving the problem. You may also notice that some students have difficulty knowing in which direction to adjust a known solution to give them a solution to a related problem (for example, for 100 – 66, some students might say 36 instead of 34). See the **Dialogue Boxes**, Students' Strategies for Solving Related Problem Sets (p. 32) and Is It 1 More or 1 Less? (p. 33), for examples of students discussing their work on problem sets.

Sessions 2, 3, and 4 Follow-Up

More Related Problem Sets After Session 2, students choose two of the related problem sets on Student Sheet 9, More Related Problem Sets, and solve them for homework. Remind students to add to the set any other problems they use to solve the final problem, and to use the 300 chart as needed.

Froggy Races 1 After Session 3, send home Student Sheet 10, Froggy Races 1, with a copy of the 300 chart.

❖ **Tip for the Linguistically Diverse Classroom** Read each problem aloud before sending Student Sheet 10 home. Have limited-English-proficient students note what Freaky Frog and Frogurt Frog each look like so they can refer to these pictures when answering the questions at home.

101 to 200 Bingo After Session 4, students can take home a copy of the 101 to 200 board, a set of Numeral Cards, and a set of Tens Cards to cut out at home. They teach 101 to 200 Bingo to someone in their family. (If students are unlikely to have scissors at home, give them some time during the school day to cut out the Numeral Cards and Tens Cards, and provide envelopes to store them.) If they wish, students can take home a copy of How to Play 101 to 200 Bingo.

Multilayer Boxes for 200 Some students may be interested in finding the multilayer boxes for 200. To encourage this activity, make a list of all the one-layer boxes students have found. Tell students they can add to the list any multilayer boxes they can make. You might want to display each new one that is found for a day or two. (There are six possible multilayer boxes if orientation doesn't matter. If it matters which rectangle is on the bottom, then there are 15 possible multilayer boxes to hold 200 cubes.)

Teacher Note — *Related Problem Sets*

Related problem sets are sets of problems that help students think about what they know to solve harder problems. Each set of problems provides a variety of possible starting places for solving the final problem in the set. For example, consider this related problem set:

150 + 50 =
150 + 60 =
156 + 50 =
156 + 59 =

The first three problems suggest different ways to solve 156 + 59. For example, you could start to solve the problem by using 150 + 50 = 200, then adding 6 + 9. Another approach is to think of the 59 as 60, add 150 + 60 = 210, then add 6 and subtract 1 (for the 59 you changed to 60). Still another way to solve this problem is to begin with 156 + 50 = 206, then add 9. All of these ways work well to solve the problem; each might be chosen by different people, depending on which of the beginning steps was most comfortable for them.

As students solve addition and subtraction related problem sets in this unit, they are thinking about how to choose a starting place. They break down problems into convenient parts, solve those parts, then combine partial solutions to get their final result. This process is easiest in addition, in which you can pull apart both numbers, then add the parts in any order. When adding 156 + 59, it doesn't matter whether you add 9 + 6 first or 150 + 50 first, as long as you combine all the subtotals at the end. Subtraction is more complex. You must keep track of which number is being subtracted and which number you are subtracting from.

Many of the related problem sets on Student Sheet 7 (Investigation 2, Sessions 2–4), emphasize adding and subtracting with multiples of 10. Students can be encouraged to count by 10's to solve some of these problems. For example, to solve 170 – 56, you might count down by 10's, 160, 150, 140, 130, 120 to subtract 50 and then subtract 6. Also, help students to use convenient landmarks such as multiples of 100 to help them solve problems. To solve 172 + 40, they might think about how 170 + 30 = 200 (the 170 is taken out of the 172 and the 30 is part of 40). That leaves 2 from the 172 + 10 from the 40, which are easily added on to 200 to get the answer of 212. Keep in mind that some students prefer to use addition to solve subtraction problems. That is why addition problems are included as part of subtraction related problem sets. For example, to solve 170 – 56, you might add on to 56 like this: 56 + 4 = 60, 60 + 40 = 100, 100 + 70 = 170. To get the final result, you then total all the numbers you used to add on to 56: 4 + 40 + 70 = 114. If any of these approaches seems awkward or difficult to you, remember that different strategies will appeal to different students—and they might not be the same ones that appeal to you. All of these are based on sound knowledge of number relationships, the structure of our number system, and the operations of addition and subtraction. While students are experimenting with different approaches, encourage them to try more than one way to solve the final problem in each set.

The third page of Student Sheet 7 contains missing addend problems. Although students may have solved this kind of problem in a story problem context, they may be unfamiliar with the notation. You may need to spend some time helping students interpret notation such as 75 + ___ = 200. It may help to provide a story context such as: "Last year I had 75 marbles in my collection. This year I have 200. How many marbles did I get since last year?" or "A pet store had 75 fish. A new shipment arrived. Now the store has 200 fish. How many fish did the store receive in the shipment?"

In the related problem sets on Student Sheet 17 (Investigation 3, Sessions 3–5), students add and subtract 100 and gradually work toward solving three-digit addition and subtraction problems by breaking up the numbers in a variety of ways. Students should be encouraged to develop strategies in which they add or subtract *from left to right* dealing with the largest components of the number first. For example, in a problem such as 678 + 325, a good strategy would be the following: 600 + 300 is equal to 900; then 70 and 20 is equal to 90, so that's 990 so far; 8 + 5 is equal to 13, that gives another 10 to add onto 990 to make 1000; and then there's 3 left from the 13, so the answer is 1003.

There are other good strategies to solve this problem. For example: 678 is close to 675. If I add on the 25 from 325, that makes 700, then add the other 300 to make 1000; then add on the extra 3 from the 678, that's 1003. Or, "I know that 678 + 22 more is 700, so take 22 out of 325. Then add the 303 that is left from 325 to 700, and the answer is 1003." Notice that this last strategy makes use of a problem that's not given in the problem set for 678 + 325. As students learn to make use of their own understanding of numbers to solve such problems, they are likely to come up with strategies different from yours, ours, or ones we might predict. As long as they can reason in a way that makes sense to them and can keep track of their work with care and accuracy, they should be encouraged to use whatever strategies work for them.

You can continue to make up additional problem sets of your own. These sets are excellent for Ten-Minute Math or to have on the board when students arrive at school.

Students' Strategies for Solving Related Problem Sets

The class is discussing these related problems during the activity Discussing Related Problem Sets.

75 + ___ = 100
75 + ___ = 275
300 – 25 =
75 + ___ = 300

Nadim: Seventy-five plus 225 is equal to 300.

Why did you pick 225?

Nadim: Because 25 plus 75 is equal to 100, so if you add 75 plus something is equal to any hundred, it has to be something minus 25. Since it's 300, then it has to be 225.

Karen: I used 75 plus something equals 100, too, but I solved it differently. If you start at 75, then it's 25 more to equal 100, so then it's 200 more to get to 300. So that's 225.

Others: I did that, too.

Did anyone use a different problem from the ones in the related problem set to solve 75 plus what equals 300?

DeShane: I have a wicked easy way. You know that 75 + 200 has to be 275. Then you're only 25 away from 300, so it's 225.

So you used 75 + something = 275 to start?

DeShane: Well, I just thought it up out of my head, but I guess it's kind of like 75 plus something equals 275, but that's not what I was thinking about when I did it.

Ahmad: I did it with 25's. See, it's like you have 75 and something else to make 300. So you just count backwards by 25's.

How did you count?

Ahmad: 275, 250, 225 [he puts up one finger for each number he says]—see, that's three 25's.

Karen: That's how I did it, too. But it shouldn't be 300 – 25. It should be 300 – 25 – 25 – 25.

DIALOGUE BOX

Is It 1 More or 1 Less?

This discussion took place while students were doing the Choice Time activity on realted problem sets. The teacher builds on what comes up in the discussion by posing a new problem for students to consider.

Let's look at this related problem set. Who has a way to think out 200 minus 135 in your head? Which other problems in the set did you use?

200 – 100 =
200 – 140 =
135 + 5 =
200 – 135 =

Nick: Well, I did 200 minus 100 first because it's easy. Then you just minus 35.

So you subtracted 100 from 200 and got 100. Then how did you know how to do 100 minus 35? Who can help with that part?

B. J.: You can do minus 30, then minus 5.

Sarah: I did 200 minus 140 first, that's 60. Then you have to subtract 5 more, so it's 65.

B. J., you look puzzled. Do you want to ask Sarah something?

B. J.: You said 200 minus 140 is 60, right? Then you said subtract 5 more, so that would be 55. But I know it's 65.

Sarah: No, but it's 5 more that you need for the answer. You only subtracted 140, so you didn't subtract enough yet. So when you had 60, it wasn't big enough.

This can be confusing in subtraction. I see quite a few people looking sort of puzzled. But is everyone pretty convinced that 200 minus 135 is 65? Yes? OK, if we're pretty convinced of that, let's see if we can use that answer to solve a new problem, 200 – 136 *[writes 200 – 136 on board underneath 200 – 135 = 65].* **Talk about this problem with someone next to you for a couple of minutes. See if you can use what we know about 200 minus 135.** *[After about 2 minutes, the teacher asks for students' solutions.]*

Rebecca: Yes—136 is 1 more than 135 so you have to take away one more from 65, so it's 64.

David: Wait, no, it's one more. I got 66.

Rebecca: It can't be 66, because if you check it, 136 plus 64 is equal to 200.

What do other people think? Is it 66 or 64?

Students: 64 . . . 66 . . . 66 . . . 64 . . .

This is a hard question. We know that 200 minus 135 is equal to 65. Everyone agrees? OK, so how can you use that to figure out 200 minus 136? Who can explain it?

Irena: Since 136 is 1 more than 135, that answer will be 1 lower because you're subtracting 1 more.

Kim: Let's say it was 134. In 200 minus 135, you have to subtract more numbers, but with 134 it's 1 less, so the answer is 1 more.

Kyle: [*comes up to the board and points*] You have to subtract 1 more here [*in 200 – 136*], so your answer should be one more here [*200 – 135*] because you're not subtracting as much, so you have a bigger answer.

Lina: I've never heard of minusing and plussing in the same problem.

Joey: I do it on the 300 chart when I get mixed up.

Can you show us on the overhead how you do it?

Joey: Look—you're going from 135 to 200, that's 65, right [*demonstrates by putting chips on 135 and 200 on a transparency of the 300 chart*]? But you want to go from 136 to 200 [*moves chip from 135 to 136*], so you have to go 1 less because you're *closer* to 200 now, see? So the answer has to be less.

Session 5

Solving Problems in the Hundreds

Materials

- Cubes, 100 charts, 300 charts, calculators, graph paper remain available
- Student Sheet 11 (1 per student)
- Crayons or markers
- Student Sheet 12 (1 per student, homework)

What Happens

Students work on one or two word problems as an assessment of their work with 100, 200, and 300 and the factors of these numbers. For the remainder of the session, students continue with the choices from Sessions 2 through 4. Their work focuses on:

- using knowledge of the factors of 100, 200, and 300 to solve problems
- communicating about their strategies with words, pictures, and numbers

Ten-Minute Math: What Is Likely? Continue to do the activity What Is Likely? in any ten-minute period you have outside of math class. Change the proportions of colors of objects in the bowls. If students understand the first set of activities, have them fill the bowl themselves to be likely to reach a particular goal. For full directions and variations, see p. 59.

Activity

Assessment

Problems with Landmarks in the Hundreds

Give a copy of Student Sheet 11, Two Problems, to each student. Cubes, calculators, 100 charts, 300 charts, and graph paper should be readily available. Ask students to choose one of the problems, to solve it in one way, to double-check their solution by solving it in a second way, and to write down their solution. Students are often good at selecting a problem of appropriate difficulty, but you may want to help some of them select their problem. You may want some students to do both problems.

❖ **Tip for the Linguistically Diverse Classroom** As you read each problem aloud, show examples, draw sketches, and/or act out ideas to make sure limited-English-proficient students understand what is being said. For example, have four students come to the front of the classroom to act out being a clean-up team.

This assessment stresses not only finding the solution to the problem but also communicating clearly about the strategies used with words, numbers, or pictures. As students finish writing their explanations, read what they have written and ask clarifying questions. Do not hesitate to insist that students add to their writing if they have not explained what they did clearly enough so someone reading their solution can understand it. Also, you may want to ask students to write an equation for the problem and solution using standard notation (students might use either multiplication or division notation).

As you look over students' work, consider the following questions:

- Is the student fluent in counting by factors of 100?
- Can the student keep track of his or her counting?
- Can the student find the solution accurately?
- Can the student explain his or her strategy clearly using words and/or pictures?
- Can the student use standard division or multiplication notation to express the problem and its solution?

For the remainder of the session, students can continue working on the choices from Sessions 2 through 4.

Session 5 Follow-Up

Another Problem in Two Ways Students solve another problem in two ways on Student Sheet 12. Make 300 charts and graph paper available to students and encourage them to use a calculator or counters at home if they are available. Students record their work so that someone reading their solutions can understand what they did. As you go over the homework later, you may want to ask students to write an equation for the problem and solution.

How Much Is 1000?

What Happens

Session 1: Numbers to 1000 Students make a book of the numbers 1 to 1000 by using ten blank 100 charts. They decide what numbers to fill in on each page so they can locate any number up to 1000 easily. When their books are completed, they work on finding the location of particular numbers in their books.

Session 2: Moving Around in the 1000 Book Students find factors of 1000 and begin a class list of these factors. They count to 1000 by 100's, 50's, and 25's, then work on finding the difference between various numbers and 1000, using these lists as aids.

Sessions 3, 4, and 5: Estimating, Adding, and Subtracting to 1000 Students choose among several activities: related problem sets that focus on adding and subtracting in the hundreds; estimating the number of beans (or other small objects) in a container; and playing Close to 1000, in which they try to make two three-digit numbers with a sum close to 1000. They compare their estimation strategies and results in a whole-group discussion. As an assessment, students develop their own problem sets for an addition problem and a subtraction problem.

Mathematical Emphasis

- Reading and writing numbers to 1000
- Locating numbers in sequence to 1000
- Getting a sense of the magnitude of multiples of 100 up to 1000
- Identifying and using important landmarks up to 1000, including the factors of 1000 and multiples of those factors (for example, 25, 50, 75, 100, 125, 150, 175, 200, . . .)
- Developing strategies for adding and subtracting numbers in the hundreds
- Estimating quantities up to 1000

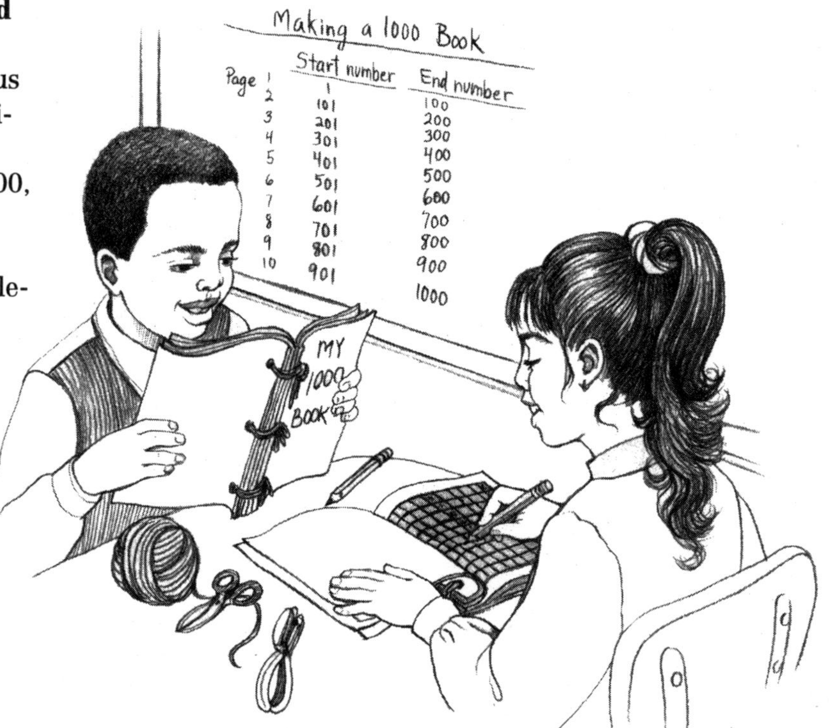

What to Plan Ahead of Time

Materials

- Construction paper or oak tag (letter-size): 2 pieces per student for making covers for 1000 books (Session 1, optional)
- Stapler or hole punch and paper fasteners or yarn for putting books together (Session 1, optional)
- Overhead projector, transparency pen (Session 1)
- Chart paper (Session 2)
- Calculators: 1 per pair (Sessions 2–5)
- Beans (or other small objects) in containers (Sessions 3–5)
- Measuring materials: spoons, scoops, paper cups (Sessions 3–5)
- Balances (Sessions 3–5, optional)
- Numeral Cards: about 6 decks for class use (use the premade cards or the sets you made in Investigation 2) and 1 deck per student for homework (Sessions 3–5)

Other Preparation

- Duplicate student sheets and teaching resources (located at the end of this unit) in the following quantities. If you have Student Activity Booklets, copy only the items marked with an asterisk.

  For Session 1

  Student Sheet 13, Find the Numbers (p. 91): 1 per pair

  Student Sheet 14, Numbers in My 1000 Book (p. 92): 1 per student (homework)

  Blank 100 charts (p. 103): 10 per student, plus extras* and a transparency

  For Session 2

  Student Sheet 15, How Far to 1000? (p. 93): 1 per student

  For Sessions 3–5

  Student Sheet 16, What's Your Estimate? (p. 94): 1 per 2–3 students

  Student Sheet 17, Related Problem Sets (p. 95): 6 of each page, cut into cards

  Student Sheet 18, Close to 100 Score Sheet (p. 97): 1 per student (optional)

  Student Sheet 19, Close to 1000 Score Sheet (p. 98): 4 per student (class), 2 per student (homework)

  Student Sheet 20, Froggy Races 2 (p. 99): 1 per student (homework)

  How to Play Close to 100 (p. 100): 1 per student (optional)

  How to Play Close to 1000 (p. 101): optional for class, 1 per student (homework)

- Prepare four containers of beans or other small objects (use the same kind in each container). If possible, use a container students can see through. Each container should have in it a quantity in the hundreds, and each should differ from the others in quantity. We suggest the following quantities: about 200, about 500, about 800, and about 1000. (Don't count these out exactly—just estimate.) Label each container with a number from 1 to 4. (Sessions 3–5)
- Make about six copies of the two pages of Student Sheet 17, and cut each page into four cards, each card with one problem set. Students work on one problem set at a time, copying it into their notebooks or onto paper.

Session 1

Numbers to 1000

Materials

- Blank 100 charts (10 per student and some extras)
- Transparency of blank 100 chart
- Construction paper or oak tag (2 pieces per student, optional)
- Stapler or hole punch and paper fasteners or yarn (optional)
- Student Sheet 13 (1 per pair)
- Student Sheet 14 (1 per student, homework)
- Overhead projector and transparency pen

What Happens

Students make a book of the numbers 1 to 1000 by using ten blank 100 charts. They decide what numbers to fill in on each page so they can locate any number up to 1000 easily. When their books are completed, they work on finding the location of particular numbers in their books. Students' work focuses on:

- writing and reading numbers to 1000
- sequencing numbers to 1000

Activity

Making a 1000 Book

Today you'll be making a book with 1000 squares in it to give us an idea of how much 1000 is. We'll be using it to count and solve problems with large numbers. I'm going to be giving each of you some blank pages like this [*show a transparency of a blank 100 chart*] to make your books. This page is just like a 100 chart without the numbers. It has 100 squares. How many of these do you think each person will need to make a book with 1000 squares? How do you know?

Have a brief discussion of the number of charts each student needs. After students have agreed on how many they need, continue with the discussion. (If they don't agree, you can leave this as an open question for students to answer as they work on their 1000 books.)

What numbers will the first page start and end with? [*1, 100*] **What about the second page?** [*101, 200*] **the third page?** [*201, 300*]

You may want to make a list on the board of the starting and ending numbers for each page to help students as they work on their books.

Your job will be to make a book of 1000 so you can find the square for any number from 1 to 1000. Some of you might want to write every single number from 1 to 1000, but that could take a long time. Can you think of any way you could write some numbers on each chart, but not all of them, so you could still find the place for any number in the whole book?

Students will have a variety of suggestions—write the first and last number on each page, write all the multiples of ten, write a few numbers in each row. See the **Dialogue Box**, Put One in Each Corner and One in the Middle (p. 40), for examples of fourth graders' ideas about this. Demonstrate students' ideas on the blank 100 chart transparency. For each idea you demonstrate, ask students whether they would be able to find other numbers on the chart easily. Each student can make his or her own decision about how to number the charts.

Students work on their 1000 books. Give out a few blank charts at a time to each student. Urge them to write their numbers lightly at first, then to double-check each page with at least two other students before they write the numbers more permanently. When all ten pages are finished, students put them together in whatever way you have decided (staple, paper fasteners, yarn); they may want to make a cover for their books as well.

Note: After this unit is over, you may want to keep students' 1000 books in a box for them to use as references throughout the year.

Activity

Teacher Checkpoint

Find the Numbers

As students finish their 1000 books, give each pair a copy of Student Sheet 13, Find the Numbers. This sheet gives students a list of numbers to locate and write in their 1000 books. As you watch students work on this list, note whether they are comfortable locating and writing numbers from 1 to 1000. If you are unsure of any student's understanding, ask him or her to show you how to locate several numbers while you are watching and to explain to you how he or she is doing this. If your students are uncertain about the order of numbers to 1000, repeat this activity with the whole class during the next few sessions, or make up more Find the Numbers lists for students to work on during Choice Time in Sessions 3 to 5.

Session 1 Follow-Up

Numbers in My 1000 Book For homework, students share their 1000 books with someone at home, and explain their numbering system. Then they find three numbers that someone else chooses, and write about how they found them on Student Sheet 14. Remind students to bring their 1000 books back to school the next day.

DIALOGUE BOX

Put One in Each Corner and One in the Middle

This discussion takes place as students are doing the activity Making a 1000 Book. The teacher has a transparency of a blank 100 chart on the overhead. The class is using the page from 301 to 400 as an example to decide what numbers they will fill in on the charts in their 1000 books. As the discussion progresses, the teacher fills in the numbers the students suggest.

We decided we didn't want to put in all the numbers. Our hands would get too tired. Did anyone have an idea about what you could do?

Emilio: Yeah, just go by tens. Do the ones across the top—1, 2, 3, 4, 5, up to 10—and then do the ones down the side—10, 20, 30, 40, up to 100.

But we're doing the chart that starts at 301 right now, so which numbers would you use?

Emilio: I'd use 301, 302, 303, all the way across to 310, then down to 320, 330, and all the way down to 300.

Vanessa: It's 400. It goes 380, 390, then 400.

Emilio: Oh, right.

[Later]

Shiro: My way is short and simple—just go down the left side—301, 311, 321, 331—then you can find any number 'cause like if it's in the 20's, you just count over from 321.

Jesse: Just use the top number, just 301.

Sarah: No! We need more.

Jesse: No, you don't. All you have to do is count across.

Sarah: No, it's not enough, then you have to keep counting by ones.

Rikki: Oh, oh, oh! There's an easier way to do it! Put one in each corner and one in the middle—301, then 310, then down at the bottom 391 and 400!

What should go in the middle?

Various students: 350 . . . 355. [*The teacher marks these on the overhead.*]

Kyle: There isn't any middle. There are four middle ones.

Joey: There isn't an exact one square that's the middle because if you go across the top, there's 10 squares so there's not one in the middle.

Shiro: You'd have to go to five and a half. You'd have to go across to between 305 and 306 and then go down.

Rikki: But it doesn't really matter which middle, just somewhere around the middle so it would help you find other numbers.

The discussion continues about where the middle of the chart is. Some students continue working on this problem as part of their effort to mark their charts.

301	302	303	304	305	306	307	308	309	310
									320
									330
									340
									350
									360
									370
									380
									390
									400

Moving Around in the 1000 Book

What Happens

Students find factors of 1000 and begin a class list of these factors. They count to 1000 by 100's, 50's, and 25's, then work on finding the difference between various numbers and 1000, using these lists as aids. Students' work focuses on:

- using what they know about factors and multiples to identify factors of 1000
- counting by 100's, 50's, and 25's to 1000
- developing strategies for subtracting numbers in the hundreds

Materials

- Students' 1000 books (from the previous session)
- Calculators (1 per pair)
- Student Sheet 15 (1 per student)
- Chart paper

Activity

Factors of 1000: Counting with Calculators

Who can think of some ways we could count to 1000 and land exactly on 1000? What are some of the factors of 1000?

Take a few suggestions. Then students work in pairs with their calculators for 10 to 15 minutes to find ways they can count to 1000. Pairs should jot down the factors they find.

Bring students back together, and list on chart paper the factors of 1000 they've found. Keep this list posted during the rest of the unit so students can add to it any other factors they discover.

Activity

Counting to 1000

With the class, count in unison by several of the factors of 1000 up to 1000. First count by 100's. Write the numbers in a list on chart paper as students say them (they need to count slowly so you have time to write). Then count together by 50's; make a list of these numbers. Finally, count by 25's and make a third list.

Ask students what they notice about these lists. What patterns do they see within each list? What relationships do they see among the three lists?

These lists provide some of the important landmarks from 1 to 1000. Keep these posted for students to use in the next activity. It may be useful to keep them posted during the rest of the unit as well.

If your students need practice locating numbers in their 1000 books, choose a few numbers from the lists and ask students to find them and label them in their books. You can circulate around the class quickly as they are finding these numbers to see how comfortable students are with the sequence of numbers in the hundreds.

Activity

How Far to 1000?

Give out Student Sheet 15, How Far to 1000? Students work in groups of two or three but record their work individually. They should use any materials or tools they need, including calculators and their 1000 books. As you circulate, ask students to show you how they solved the problems. Remind them that skip counting and using important landmarks, such as those in the lists, can help them solve these problems. Each group should double-check each solution by doing the problem at least two different ways. (Students may finish this sheet as one of their Choice Time activities in Sessions 3 to 5 if they do not finish it today.)

❖ **Tip for the Linguistically Diverse Classroom** Form groups composed of limited-English-proficient students and English-proficient students.

Estimating, Adding, and Subtracting to 1000

What Happens

Students choose among several activities: related problem sets that focus on adding and subtracting in the hundreds; estimating the number of beans (or other small objects) in a container; and playing Close to 1000, in which they try to make two three-digit numbers with a sum close to 1000. They compare their estimation strategies and results in a whole-group discussion. As an assessment, students develop their own problem sets for an addition problem and a subtraction problem. Students' work focuses on:

- developing strategies to add and subtract numbers in the hundreds
- estimating quantities of objects in the hundreds

Ten-Minute Math: Counting Around the Class Two or three times during the next few days, do the activity Counting Around the Class. Remember, Ten-Minute Math activities are done outside of math time in any spare 10 minutes you have.

Choose a factor of 1000 to count by, let's say 25. Ask students to predict what number they'll land on if they count around the class exactly once, with each student saying the next multiple of 25. Encourage students to talk about how they could figure this out without doing the actual counting.

Then start the count: the first student says "25," the next "50," the next "75," and so forth. Students can refer to the chart of counting by 25's you made in the previous session.

Stop two or three times during the count to ask a question like this:

We're at 450 now—how many students have counted so far?

After counting around once, compare the actual ending number with their predictions.

For variations on this activity, see p. 61.

Materials

- Containers of beans (or other small objects)
- Measuring materials: scoops, paper cups, spoons
- Balances (optional)
- Calculators (1 per pair)
- Student Sheet 16 (1 per 2–3 students)
- Student Sheet 17, (6 of each page, cut into cards)
- 1000 books (from previous sessions)
- Student Sheet 18 (1 per student, optional)
- Student Sheet 19 (4 per student, class; 2 per student, homework)
- Student Sheet 20 (1 per student, homework)
- Numeral Cards (about 6 decks, class; 1 deck per student, homework)
- How to Play Close to 100 (1 per student, optional)
- How to Play Close to 1000 (1 per student, optional for class; 1 per student, homework)

Activity

Choice Time: Working with Hundreds

Three Choices During most of Sessions 3, 4, and 5, students work largely independently on choice activities. They are already familiar with related problem sets, which will be one of the choices. The other two are How Many in the Jar? and Close to 1000. These are explained below. We recommend that you start with the first two choices during Session 3, then teach Close to 1000 to small groups during Session 4 (or perhaps to the whole class at the beginning of Session 4, especially if they are already familiar with Close to 100). During Session 4, have the discussion about looking at estimates (see the next activity). At the end of Session 5, leave about half of the session for students to work on the assessment activity Make Your Own Related Problem Set.

Students can probably work on two activities during each session. They may do an activity more than once. If you set up your choices at stations, show students what they will find at each station. Otherwise make sure they know where they should get all their materials.

Choice 1: How Many in the Jar?—the numbered containers of small objects you have prepared, a collection of tools for measuring (teaspoons, tiny paper cups, coffee measures, balances), copies of Student Sheet 16, calculators

Choice 2: Related Problem Sets—copies of Student Sheet 17 (pages 1–2, cut into cards), 1000 books

Choice 3: Close to 1000—copies of Student Sheet 18 (optional), Student Sheet 19, Numeral Cards, calculators, How to Play Close to 100, How to Play Close to 1000

Choice 1: How Many in the Jar?

Students work in groups of two or three. They select one of the containers of small objects. Before opening the container, they estimate "by eye" the number of objects in the container to the nearest multiple of 100. They record the number of the container and their initial estimate on Student Sheet 16, What's Your Estimate? Then they try to make a closer estimate by any method they choose, except counting *every* item one by one. They can open the container, take out the beans, and use measuring tools such as small cups, spoons, or a balance. Some students might pour a layer into a box lid, count the layer, then see how many layers are in the container. Others might use handfuls—counting the number of beans in one handful, then counting the number of handfuls in the container. They may find a calculator useful. See the **Teacher Note**, Estimating Large Quantities (p. 49), for examples of methods students have used. Students enjoy knowing that you also don't know exactly how many objects are in each container.

Choice 2: Related Problem Sets

The related problem sets on Student Sheet 17 give students practice moving around in the hundreds. These problems focus on adding and subtracting 100 and multiples of 100 as students pull apart three-digit numbers into hundreds, tens, and ones. For example, to solve 228 + 302, many students will start by adding 200 and 300. Encourage students to use their 1000 books to help them solve these problems. As in Investigation 2 Choice Time, students work on one problem at a time, writing the problem and their solution on paper or in their math notebooks and writing about how they solved the last problem in the set.

If students get confused about whether to add or subtract from an initial estimate (similar to the problem that occurs in the **Dialogue Box**, Is It 1 More or 1 Less? [p. 33]), encourage them to "act out" the problem in their 1000 book. For example, if a student is working on 989 – 190 and has already figured out that 989 – 89 – 100 = 800, but isn't sure if the answer should be one less or one more than 800, say something like: "Show me how this would work in your 1000 book. OK, you're starting at 989, now you jumped back 89 to 900, so you subtracted 89 so far. Then you jumped back 100 to 800. So how much have you subtracted so far? OK, you've subtracted 189 and you're at 800. Do you need to subtract 1 more or 1 less?"

Choice 3: Close to 1000

If your students have already done the introductory unit, *Mathematical Thinking at Grade 4*, they have played the basic game Close to 100. In this case, they will easily learn the variation Close to 1000. The only differences are (1) the goal is to make a sum as close to 1000 as possible and (2) you deal out eight Numeral Cards for each person's first turn, the player uses six and keeps two cards at each turn and receives six new cards for the next turn. You might want to start students who have never played the game with Close to 100 (use Student Sheet 18, Close to 100 Score Sheet) before they play Close to 1000. Full directions for both versions are found in the blackline masters in the back of this book for you to duplicate and post or give to students. (See the **Teacher Note**, Playing Close to 100, p. 50, for a sample game.) Calculators should be available for students to use while they play.

Activity

Looking at Our Estimates

Sometime during Session 4, when every student has had a chance to do some estimating with the containers of small objects, discuss one or two of the containers. Collect all the estimates for each of the containers you select and list them on the board. Focus the discussion around questions such as the following:

What were your first estimates when you just looked at the containers? Were some too high? Were some too low?

How did you get closer estimates? Do you feel pretty sure about your estimates? Why or why not? What could you do to get closer?

Are all the estimates we have for this container pretty close to each other? If not, why do you think they're not close? About how far apart are our estimates? About how close do you think two good estimates would be to each other? Is 50 away reasonable? Is 100 away reasonable?

Suggest that students compare the order of containers "by eye" with the order of their numerical estimates.

Which container looks like it has the least number of beans [*or other small object you have used*]**? Which has the next largest quantity? Which comes next? Which has the most?**

Once the class has agreed on an order, ask:

Does the order of our estimates match the order of the containers? Our estimates say there's about [*twice or whatever appropriate comparison*] **as many in container 4 as in container 1. Does that look about right?**

Keep the list of estimates posted so students who make new estimates can add theirs to the list. If there are very discrepant results for one of the containers, ask some other students to work with this same container during Session 5 and/or ask the students whose result was very different from the others to try their estimate again.

Activity

Assessment

Make Your Own Related Problem Set

Give each student two problems, one addition and one subtraction. Here are two suggested problems, but feel free to vary these problems in difficulty for individual students:

676 + 126 =   870 − 125 =

For each problem, students are to develop a set of related problems they can use to solve the problem, then solve each problem in the set including the original problem you gave them. As students work, circulate and ask them to explain to you how they are thinking about the problems. As you listen to what students say and look at their work, consider the following questions:

- Can students use known relationships to help them with three-digit addition and subtraction problems?
- Are they using landmark numbers such as multiples of 10, 25, or 100 to help them solve these problems?
- Can students figure out how to alter the solution of a known problem to help them with a more difficult problem? For example, if they know that 675 + 125 is equal to 800, do they know how to change this solution for the problem 676 + 126?

Sessions 3, 4, and 5 Follow-Up

Froggy Races 2 After Session 4, send home Student Sheet 20, Froggy Races 2.

Close to 1000 After Session 5, send home two copies of Student Sheet 19, Close to 1000 Score Sheet, and one copy of How to Play Close to 1000. Be sure they also have their Numeral Card deck from 101 to 200 Bingo. Students teach the game to someone at home and play several rounds. Remind students to save game materials at home for playing again later.

❖ **Tip for the Linguistically Diverse Classroom** As you read Student Sheet 20 aloud, make use of pointing and modeling actions whenever necessary to ensure comprehension.

Extensions

Finding 1000 Things Challenge students to find a group of something in the classroom or at home that is very close to 1000. These are the rules for this search:

The things have to be all the same kind of thing. For example, you can't say there are 100 people plus 100 chairs plus 500 books plus 300 pencils.

The things can't be just part of something that's much bigger. For example, you can't say there are 1000 blades of grass somewhere in the schoolyard.

You have to be able to explain how you figured out that there are close to 1000 things.

Make it clear it will be very hard to get exactly 1000. The closest anybody might come might be 980 or 1050 or even 800.

Examples of what students have found in their classroom include tiles on one wall in a hall, volumes of multivolume encyclopedias in the school, number of interlocking cubes in the class, and squares on an exhibit of graphs.

Counting Around the Class: What Would Get Us Closest to 1000? Here is a problem students can work on independently with a calculator and their 1000 books. Or you can solve it by discussing it as a whole class and trying out students' ideas.

If we counted just once around the class by some number, what number could we count by to get *closest* to 1000?

Encourage students to use the lists of counting by 100, 50, and 25 you made in Session 2 to get an initial estimate: How many 100's does it take to get to 1000? how many 50's? how many 25's? If there are 28 students in class, would we land on a number close to 1000 if we counted by 50's? Would you need to count by a bigger number or a smaller number?

Estimating Large Quantities

Teacher Note

In one classroom, students came up with a variety of methods for estimating the number of beans in the containers:

- **Layers** One method is to try to count or estimate the number of objects in a "layer" of the container—for example, the number covering the bottom. Then students can count or estimate the number of layers and multiply the number of layers by the number in each layer. Alex, Kim, and Joey used a box lid to create their own layer measure. They filled the lid with one layer of beans and counted 190. They then made layers with the rest of the beans. They made 5 full layers and about half another layer, then multiplied 190 by 5 on the calculator to get 950, mentally calculated half of 190 to be half of 100 plus half of 90 (95), and added to get 1045.
- **Scoops** Another method is to figure out how many of the objects are contained in a scoop—a teaspoon, small paper cup, or the like—then figure out how many scoops are in the entire container. Rikki, B. J., and Sarah started by using a teaspoon but decided it would take them too long and switched to using a small paper cup.
- **Using a Balance** If you have a balance in your room, students might use this tool to make their estimate. Nick and Kenyana put a handful of beans on one side of the balance and counted it. They got 178. Then they balanced that handful by pouring beans from the container onto the other side of the balance. They removed the poured beans and repeated this balancing process until they ended up with 4 groups (the original handful plus 3 groups that balanced it) and 18 extra beans. They decided 180 was close enough to 178 and 20 was close to 18, so they multiplied 180 by 4 to get 720, then added 20 to get 740.
- **Grouping** Some students may count and group the beans. Nadim, Lina Li, and Emilio counted out groups of 25 from the original container into another container. Each time Nadim put in 25, Lina Li wrote down 25 on a piece of paper. After they had counted out all the 25's, Emilio put a mark after every four 25's. Then they counted up the 100's and got 525 with 3 leftover beans for a total of 528. Note that this grouping method results in an exact count, not an estimate. You may want to discuss the difference with your students. Which estimation methods came pretty close to the actual count?

Teacher Note Playing Close to 100

The Basic Game

Here is a sample of two children playing Close to 100, using the basic scoring.

Round 1

Alex is dealt:	5	8	6	9	2	7
Lina Li is dealt:	9	1	5	5	4	7

Alex makes 58 + 29. Lina Li makes 45 + 57.

Round 2

Alex has 6 and 7 left from Round 1 and is dealt:	3	6	9	2
Lina Li has 9 and 1 left from Round 1 and is dealt:	8	2	5	0

Alex makes 36 + 62. Lina Li makes 98 + 02.

Note: Both Alex and Lina Li could have gotten closer to 100 in round 1, and Alex could have gotten closer to 100 in round 2. Can you see how?

Alex's complete game went like this:

Close to 100 Score Sheet

Name Alex — Score

Game 1

		Score
Round 1:	58 + 29 = 87	13
Round 2:	36 + 62 = 98	2
Round 3:	93 + 06 = 99	1
Round 4:	70 + 30 = 100	0
Round 5:	87 + 11 = 98	2
	Total Score:	18

Lina Li's complete game went like this:

Close to 100 Score Sheet

Name Lina Li — Score

Game 1

		Score
Round 1:	45 + 57 = 102	2
Round 2:	98 + 02 = 100	0
Round 3:	62 + 51 = 113	13
Round 4:	47 + 49 = 96	4
Round 5:	85 + 06 = 91	9
	Total Score:	28

Continued on next page

Playing with Negative and Positive Integers

Note: Students should be very comfortable with the basic game before trying this variation.

In this variation, students score the game Close to 100 using negative and positive integers. If you score 103, the difference from 100 is +3, so that is your score. If you score 98, the difference from 100 is –2, so that is your score. So, for example, the score sheets from the sample game on page 1 would look like these sample score sheets:

Close to 100 Score Sheet

Name: Alex

Game 1

		Score
Round 1:	58 + 29 = 87	-13
Round 2:	36 + 62 = 98	-2
Round 3:	93 + 06 = 99	-1
Round 4:	70 + 30 = 100	0
Round 5:	87 + 11 = 98	-2
	Total Score:	-18

Close to 100 Score Sheet

Name: Lina Li

Game 1

		Score
Round 1:	45 + 57 = 102	+2
Round 2:	98 + 02 = 100	0
Round 3:	62 + 51 = 113	+13
Round 4:	47 + 49 = 96	-4
Round 5:	85 + 06 = 91	-9
	Total Score:	+2

The player with the total score *closest to 0* wins. In this case, +2 is 2 away from 0 and –18 is 18 away from 0, so Lina Li wins.

Scoring this way changes the strategy for the game. Even though Alex got four scores very close to 100, he did not compensate for his negative values with some positive ones. Lina Li had some totals further away from 100, but she balanced her negative and positive scores more evenly to come out with a total score closer to zero.

INVESTIGATION 4

Making a 10,000 Chart

What Happens

Sessions 1, 2, and 3: 10,000 Squares on the Wall The class makes a 10,000 chart out of one hundred 100 charts. As they did for the 1000 books, they decide which numbers are necessary on each chart. Students use calculators to discuss what numbers come next and practice reading and writing numbers in the thousands as they make the chart. When the chart is completed, students locate numbers on the chart and add multiples of 100 to and subtract multiples of 100 from numbers on the chart.

Mathematical Emphasis

- Reading, writing, and sequencing numbers in the thousands
- Getting a sense of the magnitude of 10,000
- Understanding the structure of 10,000 (for example, that it can be constructed of 10 thousands or 100 hundreds)
- Adding multiples of 100 to and subtracting multiples of 100 from numbers in the thousands

What to Plan Ahead of Time

Materials

- Calculators: 1 per pair (Sessions 1–3)
- Transparency of blank 100 chart (made in Investigation 3) (Sessions 1–3)

Other Preparation

- Duplicate student sheets and teaching resources (located at the end of this unit) in the following quantities. If you have Student Activity Booklets, copy only the items marked with an asterisk.

  For Sessions 1–3

  Blank 100 charts (p. 103): about 120 (100 for the 10,000 chart and extras* in case students need to make rough drafts)

  Student Sheet 19, Close to 1000 Score Sheet (p. 98): 2 per student (homework)

- Clear a space in your room for the 10,000 display. You'll be putting up the 100 charts in a 10 by 10 array. You'll need a space on the wall about 8 feet wide and 10 feet high to hang this array. Some teachers have put the chart out in the hall, but it is easier to work with students on locating numbers if it is in the room.

Sessions 1, 2, and 3

10,000 Squares on the Wall

Materials

- Blank 100 charts (100 and some extras)
- Transparency of blank 100 chart
- Calculators (1 per pair)
- Student Sheet 19 (2 per student, homework)

What Happens

The class makes a 10,000 chart out of one hundred 100 charts. As they did for the 1000 books, they decide which numbers are necessary on each chart. Students use calculators to discuss what numbers come next and practice reading and writing numbers in the thousands as they make the chart. When the chart is completed, students locate numbers on the chart and add multiples of 100 to and subtract multiples of 100 from numbers on the chart. Their work focuses on:

- reading and writing numbers in the thousands
- understanding the sequence of numbers in the thousands
- adding multiples of 100 to and subtracting multiples of 100 from numbers in the thousands

Ten-Minute Math: Counting Around the Class During the next few days, continue to do the activity Counting Around the Class in any spare ten minutes you have outside of math class. Count by larger numbers, such as 100, 500, 1000, and 2000. Stop several times during the count to ask questions such as:

How many students have counted so far? What number do you think we will end on now?

For full directions and variations, see p. 61.

Activity

100 for Every Student

Introducing the 10,000 Chart Activity Tell students that the class is going to make a giant 10,000 chart:

Now that you've made 1000, we're going to make *10,000*. Do you think it's possible for us to show 10,000? How much space do you think we'd need? How long would it take to make a 10,000 chart?

Have a few minutes of preliminary discussion about what 10,000 squares might look like. Students may point out they already have made more than 10,000 if they put all their 1000 books together. This would be a good time for students to figure out how many squares they have altogether in the class's 1000 books.

How Many Hundreds Will We Need? Show the blank 100 chart transparency on the overhead projector.

We're going to make our 10,000 chart out of these blank 100 charts just like we made our 1000 books. My first question is: How many of these charts will we need to make 10,000?

Students work in groups of two or three on this question for a few minutes. Discuss with students how they thought about this problem. Ask them to explain their reasoning. If there is not a consensus about how many 100 charts will be needed by the end of this discussion, leave this as an open question. It will be resolved through the subsequent activities.

Now we're going to start making our 10,000 chart. If everybody in the class filled in the numbers on one of these blank 100 charts, how far would we get? For example, let's say Shoshana took one chart and wrote 1 to 100, then David took one chart and wrote the next hundred numbers, and Marci took a chart and wrote out the next hundred, and everyone in the class did one—how far would we get?

Encourage students to give their opinions. Then count around the classroom once by 100's to see where you end (with the first student saying 100, the next 200, the next 300, and so forth).

Next, explain how each student will be responsible for one of the "hundreds." The first student will write 1 to 100, the next student will write 101 to 200, and so forth. On the board or overhead, list students' names. Have the class help you make decisions about who will write which numbers:

If Marci ends with 300, what number should Rebecca start with? What number will she end with? If Rafael started with 1201, what number will Teresa, who's two hundreds away, end with?

You can skip around on the list to make the questions more challenging. If students aren't sure what comes next, they can use their calculators to find out. For example, in one classroom, students weren't sure what came after 1100. Students entered 1100 on their calculators, then added one more. At the end of the discussion, you'll end up with a list that shows which numbers each student will write on his or her 100 chart:

Shoshana	1 – 100
David	101 – 200
Marci	201 – 300
Rebecca	301 – 400
Tuong	2301 – 2400

As it comes up, you can point out how people say many of these numbers in different ways. For example, 2301 might be read as "twenty-three oh one," "twenty-three hundred and one," or "two thousand, three hundred, one."

Now hand each student a blank 100 chart. He or she writes 100 numbers on his or her chart: The first student writes 1 to 100, the next student writes 101 to 200, the next student writes 201 to 300, and so forth, according to the list you have made.

Note: Before the next activity, post the students' charts in order, leaving enough blank space for other 100 charts that complete 10,000. So, at this point, if you have 24 students in your class, you'll have twenty-four 100 charts with the numbers up to 2400. Post these in two rows of ten 100 charts and one row of the four remaining 100 charts, with space for another seventy-six 100 charts.

Activity

Completing the 10,000 Chart

Now we're going to put up more hundreds until we have 10,000. How many more 100 charts do you think we'll need? How do you know?

After a few minutes of discussion, you can go on even if this question isn't resolved for sure. If there are differences of opinion, you can tell students they're actually going to put up the 100 charts needed to get to 10,000, so that they'll soon find out.

We do not need to write in every single number up to 10,000. Just as we did for the 1000 books, we can put just enough numbers on each chart from here on so we can find any number we want. How do you think we should do it?

This time students can use their experience with the 1000 books to talk about what worked for them. What numbers did they need on each chart so they could find any number? Let students decide how to mark each of the remaining 100 charts. After a decision has been reached, assign charts to individual students or to groups. For example, if your class is working in groups of four, each group of four may be responsible for 12 or 13 of the hundreds. So the first group of four might be responsible for the 2500 chart (the one that goes from 2401 to 2500), the 2600 chart (the one that goes from 2501 to 2600), all the way up to the 3600 chart (the one that goes from 3501 to 3600). Another strategy is to give each group only four or five at a time to work on. When they have completed those, you can give them their next batch of four or five.

Note: We have found it useful to establish the convention that each 100 chart is named by the *last* number on the chart. So, we call the chart that goes from 2401 to 2500 "the twenty-five hundred chart" or the "two thousand five hundred chart," and we call the chart that goes from 5701 to 5800 "the fifty-eight hundred chart" or "the five thousand eight hundred chart." This is just a convenient way of knowing we are all talking about the same chart. If your students come up with a better way to name their charts, that's fine.

Figuring out which numbers to write and where they go will be difficult for some students as they move further into the thousands. Circulate among your students and help them think about how to write the numbers and which spaces they go in. Encourage them to help one another and to look at some of the 100 charts that are already completed when they are stuck.

Also, encourage students to use the calculator to help them. Suppose, for example, a student wasn't sure how to write the first number of her 100 chart. She could use the calculator to add 1 to the last number of the previous 100 chart. So if the last 100 chart ended with 2400, she can put into the calculator: [2] [4] [0] [0] [+] [1] [=]. The calculator display will show her 2401. If she wants to, she can keep going, adding ones on the calculator, until she gets to the next number she needs to write. Remind students how to use the calculator to count by a particular number (for example, [+] [1] [=] [=] [=] [=] or [+] [10] [=] [=] [=] [=]).

As 100 charts are completed, put them up in the right place on your 10,000 display.

Activity

What Can We See on the 10,000 Chart?

When the 10,000 chart is completed, ask students to look at the whole thing and have a discussion focusing on the structure of the whole chart:

How many numbers are on each 100 chart? How many numbers are in one whole row of the 10,000 chart? How many hundreds are in one row of the 10,000 chart? What can you tell me about the numbers in this row? What can you tell me about the numbers on this 100 chart? How many hundreds are in half the 10,000 chart? How many numbers are in half the 10,000 chart?

Ask students to come up and locate numbers on the chart:

Where is the 200 chart? Where is the 500 chart? Which one is the 1000 chart? Where is the 2500 chart? Which chart is this? Which chart is this? How do you know?

Help students find different charts by counting up from the beginning by one hundreds (100, 200, 300, . . .) or by counting from a chart that they recognize:

What chart that you know could help you find the 7200 square?

Some students might reply: "The thousands are at the end of a row, like the last chart in the top row is 1000, then the one under that is 2000, then 3000. So we could count down by thousands until we got to 7000, then two more is 7200."

Other students might suggest: "All the ones with 200 in them are the second ones in. Like 200 is the second one in the top row, then 1200 in the next row, so 7200 is in the seventh row."

Once students are comfortable with finding numbers, add questions that require them to add and subtract multiples of 100:

Who'd like to come up and find 2500? Who can find 8700? Who can find a number 100 less than 4500? Find 3000; now find a number 400 more than 3000.

Many teachers continue using this activity as a Ten-Minute Math activity.

Activity

Choosing Student Work to Save

As the unit ends, you may want to use one of the following options for creating a record of students' work on this unit:

- Students look back through their folders or notebooks and write about what they learned in this unit, what they remember most, and what was hard or easy for them. You might have students do this work during their writing time.
- Students select one or two pieces of their work as their best, and you also choose one or two pieces of their work to be saved. This work is saved in a portfolio for the year. You might include students' written solutions to the Assessment, Problems with Landmarks in the Hundreds, Investigation 2, Session 5, and any other assessment tasks from this unit. Students can create a separate page with brief comments describing each piece of work.
- You may want to send a selection of work home for families to see. Students write cover letters, describing their work in this unit. This work should be returned if you are keeping a year-long portfolio of mathematics work for each student.

Sessions 1, 2, and 3 Follow-Up

Close to 1000 For homework, students continue to play Close to 1000 with someone at home. Students will need additional copies of the Close to 1000 Score Sheet (Student Sheet 19), or they can use blank paper to make their own.

What Is Likely?

Basic Activity

Students make judgments about drawing objects of two different colors from a clear container. They first decide whether it's likely they'll get more of one color or the other. Then they draw out ten objects, one at a time, recording the color of each and replacing that object before picking the next one. Students then discuss whether what they expected to happen did happen. They repeat the activity with another sample of ten objects.

What Is Likely? involves students in thinking about ratio and proportion and in considering the likelihood of the occurrence of a particular event. Ideas about probability are notoriously difficult for children and adults. In the early and middle elementary grades, we simply want students to examine familiar events in order to judge how likely or unlikely they are. In this activity, students focus on:

- visualizing the ratio of two colors in a collection
- making predictions and comparing predictions with outcomes
- exploring the relationship between a sample and the group of objects from which it comes

Materials

- A clear container, such as a fishbowl or large glass or clear plastic jar
- Objects that are very similar in size and shape but come in two colors (wooden cubes, beans, beads, marbles)

Procedure

Step 1. Fill the container with two colors of cubes, beads, or beans. When you first use this activity, put much more of one color into the container. For example, out of every 10 cubes you put in the container, you might use 9 red and 1 yellow. Thus, if you used 40 cubes, 36 would be red and 4 would be yellow. Mix these well inside the container. Continue to use these markedly different proportions for a while.

Step 2. Students predict which color they will get the most of if they draw 10 objects out of the container. Carry the container around the room so all students can get a good look at its contents. Then ask students to make their predictions. "What is likely to happen if we pull out 10 objects? Will we get more yellows or more reds? Will we get a lot more of one color than the other?"

Step 3. Students draw 10 objects from the container, replacing after each draw. Ask a student to close his or her eyes and draw out one object. Record its color on the board before the student puts the object back. Ask 9 more students to pick an object, then replace it after you have recorded its color. Record colors using tallies.

Red *𝍸 ///*

Yellow *//*

Step 4. Discuss what happened. "Is this about what you expected? Why or why not?" Even if you have a 9:1 ratio of the two colors, you won't always draw out a sample that is exactly 9 of one color and 1 of the other. Eight red and 2 yellow or 10 red and 0 yellow are also likely samples. Ask students whether what they got is likely or unlikely, given what they can see in the container. What would be *unlikely*, or surprising? (Of course, surprises can happen, too—just not very often!)

Step 5. Try it again. Students will probably want to try drawing another 10 objects to see what happens. "Do you still think it's likely that we'll get mostly reds again? Why? About how many do you think we'll get?" Draw objects, tally their colors, and discuss in the same way.

Variations

Different Color Mixes Try a 3:1 ratio—3 of one color for every 1 of the other color. Also try an equal amount of the two colors.

Continued on next page

Different Objects Try two colors of a different kind of object. Does a change like this affect the outcome?

The Whole Class Picks See what happens when each student in the class draws (and puts back) one object. Before you start, ask, "If all 28 of us pick an object, about how many reds do you think we'll get? Is it more likely you'll pick a red or a yellow? A little more likely or a lot more likely?"

Students Fill the Container Ask students to help you decide what proportions of each color to put in the container. Set a goal—for example:

- How can we fill the container so it's very likely we'll get mostly yellows when we draw 10?
- How can we fill the container so it's unlikely we'll get more than 1 red?
- How can we fill the container so we'll get close to the same number of reds and yellows when we draw 10?

After students decide how to fill the container, draw objects, as in the basic activity above, to see if their prediction works.

Three Colors Put an equal number of two colors (red and yellow) in the container, and mix in many more or many fewer of a third color (blue). "If 10 people pick, about how many of each color (red, yellow, and blue) do you think we will get? Do you think we'll get the same number of red and yellow, or do you think we will get more of one of them?"

Counting Around the Class

Basic Activity

Students count around the class by a particular number—that is, if counting by 2's, the first student says "2," the next student says "4," the next "6," and so forth. Before the count starts, students try to predict on what number the count will end. During and after the count, students discuss relationships between the chosen factor and its multiples.

Counting Around the Class is designed to give students practice with counting by many different numbers and to foster numerical reasoning about the relationships among factors and their multiples. Students focus on:

- becoming familiar with multiplication patterns
- relating factors to their multiples
- developing number sense about multiplication and division relationships

Materials

Calculators (for variation)

Procedure

Step 1. Choose a number to count by. For example, if the class has been working with quarters recently, you might want to count by 25's.

Step 2. Ask students to predict the target number. "If we count by 25's around the class, what number will we end on?" Encourage students to talk about how they could figure this out without doing the actual counting.

Step 3. Count around the class by your chosen number. "25 . . . 50 . . . 75 . . ." If some students seem uncertain about what comes next, you might put the numbers on the board as they count; seeing the visual patterns can help some students with the spoken pattern.

You might count around a second time by the same number, starting with a different person, so students will hear the pattern more than once and have their turns at different points in the sequence.

Step 4. Pause in the middle of the count to look back. "We're up to 375 counting by 25's. How many students have counted so far? How do you know?"

Step 5. Extend the problem. Ask questions like these:

"Which of your predictions were reasonable? Which were possible? Which were impossible?" (A student might remark, for example, "You couldn't have 510 for 25's because 25 lands only on the 25's, the 50's, the 75's, and the 100's.")

"What if we had 32 students in this class instead of 28? Then where would we end?"

"What if we used a different number? This time we counted by 25's and ended on 700; what if we counted by 50's? What number do you think we would end on? Why do you think it will be twice as big? How did you figure that out?"

Variations

Multiplication Practice Use single-digit numbers to provide practice with multiplication pairs (that is, count by 2's, 3's, 4's, 5's, 6's, and so forth). In counting by numbers other than 1, students usually first become comfortable with 2's, 5's, and 10's, which have very regular patterns. Soon they can begin to count by more difficult single-digit numbers: 3, 4, 6 and (later) 7, 8, and 9.

Landmark Numbers When students are learning about money or about our base ten system of numeration, they can count by 20's, 25's, 50's, 100's, and 1000's. Counting by multiples of 10 and 100 (30's, 40's, 600's) will support students' growing familiarity with the base ten system of numeration.

Continued on next page

Making Connections When you choose harder numbers, pick those that are related in some way to numbers students are very familiar with. For example, once students are comfortable counting by 25's, have them count by 75's. Ask students how knowing the 25's will help them count by 75's. If students are fluent with 3's, try counting by 6's or by 30's. If students are fluent with 10's and 20's, start working on 15's. If they are comfortable counting by 15's, ask them to count by 150's or 1500's.

Large Numbers Introduce large numbers—such as 2000, 5000, 1500, or 10,000—so students begin to work with combinations of these less familiar numbers.

Using the Calculator On some days you might have everyone use a calculator or have a few students use calculators to skip count while you are counting around the class. On most calculators, the [=] key provides a built-in constant function, allowing you to skip count easily. For example, if you want to skip count by 25's, you press your starting number (let's say [0]), the operation you want to use (in this case, [+]), and the number you want to count by (in this case, [25]). Then, press the [=] key each time you want to add 25. So, if you press

[0] [+] [25] [=] [=] [=] [=]

you will see on your screen 25, 50, 75, 100.

Special Notes

Letting Students Prepare When introducing an unfamiliar number to count by, students may need some preparation before they try to count around the class. Ask students to work in pairs to figure out, with whatever materials they want to use, on what number the count will end.

Avoiding Competition Be sensitive to potential embarrassment or competition that may occur if some students have difficulty figuring out their number. One teacher allowed students to volunteer for the next number, rather than counting in a particular order. Other teachers have made the count a cooperative effort, establishing an atmosphere in which students readily helped one another, and anyone felt free to ask for help.

Counting Patterns Students write out a counting pattern up to a target number (for example, by 25's up to 500). Then they write about what patterns they see in their counting. Calculators can be used for this.

Mystery Number Problems Provide an ending number and ask students to figure out what factor they would have to count by to reach it. For example: "I'm thinking of a mystery number. I figured out that if we counted around the class by my mystery number today, we would get to 2800. What is the mystery number?"

Or you might provide students with the final number and the factor and ask them to figure out the number of students in the class. "When a certain class counts by 25's, the last student says 550. How many students are in the class?" Calculators can be used.

The following activities will help ensure that this unit is comprehensible to students who are acquiring English as a second language. The suggested approach is based on *The Natural Approach: Language Acquisition in the Classroom* by Stephen D. Krashen and Tracy D. Terrell (Alemany Press, 1983). The intent is for second-language learners to acquire new vocabulary in an active, meaningful context.

Note that *acquiring* a word is different from *learning* a word. Depending on their level of proficiency, students may be able to comprehend a word on hearing it during an investigation, without being able to say it. Other students may be able to use the word orally but not read or write it. The goal is to help students naturally acquire targeted vocabulary at their present level of proficiency.

We suggest using these activities just before the related investigations. The activities can also be led by English-proficient students.

Investigations 1–4

layer, box

1. Show several different-size boxes. Then hold up one of the boxes and point out its depth.
2. Using centimeter cubes, demonstrate how many layers of cubes could be placed in this box.
3. Hold up another box. This time ask students to predict how many layers of cubes would fit inside. Have students use the cubes to prove or disprove their answers. Students can build layers inside or next to the boxes. Continue with this format until students understand the concept of *layer*.

jump, ahead, land exactly

1. After you model the meaning of the word *jump*, choose three students to participate in a jumping race across the room. As they do, become a "sports commentator," relaying to the audience who is ahead at all times.

   We can see that Rashaida is ahead in this race—No. wait! Jesse is jumping faster, faster—Jesse takes the lead and now he is ahead of everyone else!

2. After the race, ask questions that can be answered with one-word responses about who was in the lead.

   Who was ahead at the beginning of the race? Who was ahead at the end of the race?

3. Next put a piece of paper on the floor and model jumping over, across, and landing exactly on the piece of paper.
4. Choose a student to stand approximately 10 feet away from the paper. Ask the group to predict how many jumps the student will take before landing exactly on the paper. After the student has jumped, have students compare their predictions with the actual number. Vary the distances as you continue with this same format with other students.

Blackline Masters

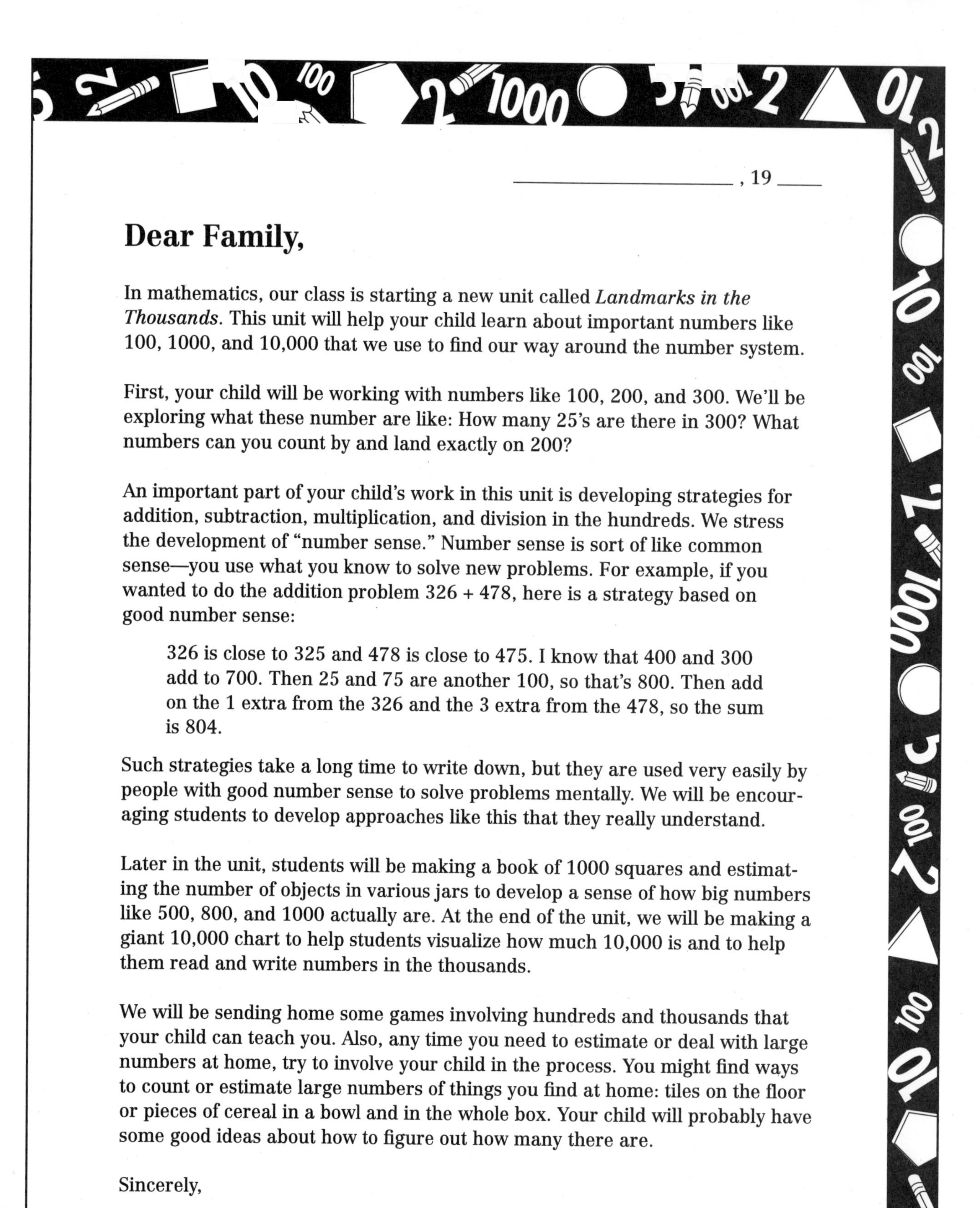

_______________, 19 ____

Dear Family,

In mathematics, our class is starting a new unit called *Landmarks in the Thousands*. This unit will help your child learn about important numbers like 100, 1000, and 10,000 that we use to find our way around the number system.

First, your child will be working with numbers like 100, 200, and 300. We'll be exploring what these number are like: How many 25's are there in 300? What numbers can you count by and land exactly on 200?

An important part of your child's work in this unit is developing strategies for addition, subtraction, multiplication, and division in the hundreds. We stress the development of "number sense." Number sense is sort of like common sense—you use what you know to solve new problems. For example, if you wanted to do the addition problem 326 + 478, here is a strategy based on good number sense:

> 326 is close to 325 and 478 is close to 475. I know that 400 and 300 add to 700. Then 25 and 75 are another 100, so that's 800. Then add on the 1 extra from the 326 and the 3 extra from the 478, so the sum is 804.

Such strategies take a long time to write down, but they are used very easily by people with good number sense to solve problems mentally. We will be encouraging students to develop approaches like this that they really understand.

Later in the unit, students will be making a book of 1000 squares and estimating the number of objects in various jars to develop a sense of how big numbers like 500, 800, and 1000 actually are. At the end of the unit, we will be making a giant 10,000 chart to help students visualize how much 10,000 is and to help them read and write numbers in the thousands.

We will be sending home some games involving hundreds and thousands that your child can teach you. Also, any time you need to estimate or deal with large numbers at home, try to involve your child in the process. You might find ways to count or estimate large numbers of things you find at home: tiles on the floor or pieces of cereal in a bowl and in the whole box. Your child will probably have some good ideas about how to figure out how many there are.

Sincerely,

Name Date

Student Sheet 1

Miniature 100 Charts

1	2	3	4	5	6	7	8	9	10
11	12	13	14	15	16	17	18	19	20
21	22	23	24	25	26	27	28	29	30
31	32	33	34	35	36	37	38	39	40
41	42	43	44	45	46	47	48	49	50
51	52	53	54	55	56	57	58	59	60
61	62	63	64	65	66	67	68	69	70
71	72	73	74	75	76	77	78	79	80
81	82	83	84	85	86	87	88	89	90
91	92	93	94	95	96	97	98	99	100

We counted by ________.

We landed on 100 exactly. (yes__ no__)

To reach 100 exactly, we made__jumps.

1	2	3	4	5	6	7	8	9	10
11	12	13	14	15	16	17	18	19	20
21	22	23	24	25	26	27	28	29	30
31	32	33	34	35	36	37	38	39	40
41	42	43	44	45	46	47	48	49	50
51	52	53	54	55	56	57	58	59	60
61	62	63	64	65	66	67	68	69	70
71	72	73	74	75	76	77	78	79	80
81	82	83	84	85	86	87	88	89	90
91	92	93	94	95	96	97	98	99	100

We counted by ________.

We landed on 100 exactly. (yes__ no__)

To reach 100 exactly, we made__jumps.

1	2	3	4	5	6	7	8	9	10
11	12	13	14	15	16	17	18	19	20
21	22	23	24	25	26	27	28	29	30
31	32	33	34	35	36	37	38	39	40
41	42	43	44	45	46	47	48	49	50
51	52	53	54	55	56	57	58	59	60
61	62	63	64	65	66	67	68	69	70
71	72	73	74	75	76	77	78	79	80
81	82	83	84	85	86	87	88	89	90
91	92	93	94	95	96	97	98	99	100

We counted by ________.

We landed on 100 exactly. (yes__ no__)

To reach 100 exactly, we made__jumps.

Name ______________________ Date ______

Student Sheet 2

Factors of 100

Use this recording sheet with the miniature 100 charts.

Numbers we tried	Did we land on 100 exactly?	If yes, how many in 100?
______	Yes No	______
______	Yes No	______
______	Yes No	______
______	Yes No	______
______	Yes No	______
______	Yes No	______
______	Yes No	______
______	Yes No	______
______	Yes No	______
______	Yes No	______
______	Yes No	______
______	Yes No	______
______	Yes No	______
______	Yes No	______
______	Yes No	______
______	Yes No	______
______	Yes No	______
______	Yes No	______
______	Yes No	______
______	Yes No	______

Name Date

Student Sheet 3

More on Factors of 100

Answer the following questions about factors of 100. Explain your answer so that someone else can understand how you solved the problem. You can use pictures, numbers, or words.

1. Is 38 a factor of 100? Explain how you know.

2. Marilyn said that 50 is the largest factor of 100 and that anything above 50 is "too big." Do you agree? Why or why not?

Name Date

How Far to 100?

Frogurt the Frog is jumping on your 100 chart. Frogurt jumps only 1 square at a time. Frogurt always counts his jumps, but he doesn't count the square he starts on. So if he started on 30 and took 2 jumps, he would land on 32. If he started on 50, it would take him 5 jumps to get to 55.

How many jumps will it take for Frogurt to get to 100 from these squares on the 100 chart?

Start at 35. How many jumps to 100?

Start at 25. How many jumps to 100?

Start at 75. How many jumps to 100?

Start at 15. How many jumps to 100?

Start at 40. How many jumps to 100?

Start at 43. How many jumps to 100?

Start at 50. How many jumps to 100?

Start at 56. How many jumps to 100?

Name Date

Student Sheet 5

Start at 11. How many jumps to 100?

Name Date

Student Sheet 6

Exploring Multiples of 100

Choose one of these factors of 100. Circle it.

4 5 10 20 25

How many of your number makes 100, 200, and 300? Fill in the chart.

You can try 400, 500, and higher hundreds if you want to.

	How many?
100	
200	
300	

I know there are ______ in 300 because ______

This is what we noticed about the patterns for our factor:

Name Date

Student Sheet 7

Jumping on the 100 Chart

120 + 10 = 120 + 20 = **120 + 30 =**	280 + 10 = 280 + 20 = **280 + 30 =**
64 + 10 = 50 + 50 = 64 + 30 = 60 + 50 = **64 + 50 =**	140 + 30 = 142 + 30 = 142 + 40 = **142 + 39 =**

Name Date

Student Sheet 7

Related Problem Sets (page 2 of 4)

172 + 10 = 172 + 30 = **172 + 40 =**	160 + 40 = 168 + 40 = 160 + 70 = **168 + 72 =**
170 – 10 = 170 – 50 = 56 + 4 = 170 – 6 = **170 – 56 =**	242 – 10 = 242 – 50 = 56 + 44 = **242 – 56 =**

Name Date

Related Problem Sets (page 3 of 4)

75 + 25 = 75 + 100 = **75 + ____ = 200**	75 + ____ = 100 75 + ____ = 275 300 – 25 = **75 + ____ = 300**
200 – 100 = 200 – 120 = 124 + 6 = **200 – 124 =**	40 + 20 = 40 + 60 = 200 – 20 = **40 + ____ = 200**

Name Date

Related Problem Sets (page 4 of 4)

200 – 100 = 200 – 140 = 135 + 5 = **200 – 135 =**	100 – 40 = 102 – 40 = 102 – 6 = 46 + 10 = **102 – 46 =**
100 + 40 = 110 + 40 = 100 + 50 = **110 + 42 =**	136 + 10 = 136 + 50 = 130 + 50 = **136 + 56 =**

Name ______________________________ Date ____________

Student Sheet 8

Factors of 300

Choose a number to count by. Pick one you think will land exactly on 300.

Skip count by this number on your calculator.

Does it work? If so, write how many of your number it takes to get to 300.

Numbers we tried	Did we land on 300 exactly?	If yes, how many in 300?
________	Yes No	________
________	Yes No	________
________	Yes No	________
________	Yes No	________
________	Yes No	________
________	Yes No	________
________	Yes No	________
________	Yes No	________
________	Yes No	________
________	Yes No	________
________	Yes No	________
________	Yes No	________
________	Yes No	________
________	Yes No	________
________	Yes No	________
________	Yes No	________
________	Yes No	________

Name Date

Student Sheet 9

More Related Problem Sets

150 – 10 = 150 – 20 = **150 – 30 =**	220 – 10 = 220 – 20 = **220 – 30 =**
73 + 10 = 50 + 50 = 73 + 20 = 70 + 50 = **73 + 50 =**	120 + 20 = 123 + 20 = 123 + 30 = **123 + 29 =**

Name Date

Student Sheet 10

Froggy Races 1

1. Two frogs had a race on the 300 chart. Freaky Frog took 6 jumps of 50. Frogurt Frog took 10 jumps of 25. Who was ahead? How do you know?

   I think ______________________ was ahead. This is how I figured this out:

2. In the second race, Freaky Frog started with 12 jumps of 20. Frogurt took 5 jumps of 50. Who was ahead? How did you figure it out?

   I think ______________________ was ahead. This is how I figured this out:

3. In the last race, Freaky took 40 jumps of 5. Frogurt decided to take jumps of 10. How many jumps of 10 did Frogurt have to take to beat Freaky? Why do you think so?

   I think Frogurt had to take ______ jumps. This is why I think so:

Name Date

Two Problems

Choose one problem to solve. Solve it in two ways. Write about how you solved it so someone else can understand what you did. Illustrate with a picture if that would help the reader.

(You can solve both problems if you have time.)

1. Chris needs to buy 300 cookies for the class party. The cookies come in bags of 25. How many bags does Chris need to buy? Write down how you thought about this problem.

2. In Kim's school, each student is on a team that does a clean-up job every week around the school. There are 4 students on each team and 240 students in the school. How many clean-up teams are there?

Name Date

Student Sheet 12

Another Problem in Two Ways

Choose one problem to solve. Solve it in two ways. Write about how you solved it so someone else can understand what you did. Illustrate with a picture if that would help the reader.

(You can solve both problems if you have time.)

1. DeShane needs to buy 400 balloons for the class party. The balloons come in bags of 25. How many bags does DeShane need to buy? Write down how you thought about this problem.

2. In Lina's school, each student is on a team that does some work in the school garden every week. There are 5 students on each team and 250 students in the school. How many teams are there?

300 CHART

1	2	3	4	5	6	7	8	9	10
11	12	13	14	15	16	17	18	19	20
21	22	23	24	25	26	27	28	29	30
31	32	33	34	35	36	37	38	39	40
41	42	43	44	45	46	47	48	49	50
51	52	53	54	55	56	57	58	59	60
61	62	63	64	65	66	67	68	69	70
71	72	73	74	75	76	77	78	79	80
81	82	83	84	85	86	87	88	89	90
91	92	93	94	95	96	97	98	99	100
101	102	103	104	105	106	107	108	109	110
111	112	113	114	115	116	117	118	119	120
121	122	123	124	125	126	127	128	129	130
131	132	133	134	135	136	137	138	139	140
141	142	143	144	145	146	147	148	149	150
151	152	153	154	155	156	157	158	159	160
161	162	163	164	165	166	167	168	169	170
171	172	173	174	175	176	177	178	179	180
181	182	183	184	185	186	187	188	189	190
191	192	193	194	195	196	197	198	199	200
201	202	203	204	205	206	207	208	209	210
211	212	213	214	215	216	217	218	219	220
221	222	223	224	225	226	227	228	229	230
231	232	233	234	235	236	237	238	239	240
241	242	243	244	245	246	247	248	249	250
251	252	253	254	255	256	257	258	259	260
261	262	263	264	265	266	267	268	269	270
271	272	273	274	275	276	277	278	279	280
281	282	283	284	285	286	287	288	289	290
291	292	293	294	295	296	297	298	299	300

ONE-CENTIMETER GRAPH PAPER

101 TO 200 BINGO BOARD

101	102	103	104	105	106	107	108	109	110
111	112	113	114	115	116	117	118	119	120
121	122	123	124	125	126	127	128	129	130
131	132	133	134	135	136	137	138	139	140
141	142	143	144	145	146	147	148	149	150
151	152	153	154	155	156	157	158	159	160
161	162	163	164	165	166	167	168	169	170
171	172	173	174	175	176	177	178	179	180
181	182	183	184	185	186	187	188	189	190
191	192	193	194	195	196	197	198	199	200

Materials

- 101 to 200 Bingo Board
- One deck of Numeral Cards
- One deck of Tens Cards
- Colored pencils, crayons, or markers

Players: 2

How to Play

1. Each player takes a 1 from the Numeral Card deck and keeps this card throughout the game.
2. Shuffle the two decks of cards. Place each deck face down on the table.
3. Players use just one Bingo Board. You will take turns and work together to get a Bingo.
4. To determine a play, draw two Numeral Cards and one Tens Card. Arrange the 1 and the two other numerals to make a number between 100 and 199. Then add or subtract the number on your Tens Card. Circle the resulting number on the 101 to 200 Bingo Board.
5. Wild Cards in the Numeral Card deck can be used for any numeral from 0 through 9. Wild Cards in the Tens Card deck can be used as + or – any multiple of 10 from 10 through 70.
6. Some combinations cannot land on the 101 to 200 Bingo Board at all. Make up your own rules about what to do when this happens. (For example, a player could take another turn, or the Tens Card could be *either* added or subtracted in this instance.)
7. The goal is for the players together to circle five adjacent numbers in a row, in a column, or on a diagonal. Five circled numbers is a Bingo.

0	0	1	1
0	0	1	1
2	2	3	3
2	2	3	3

4	4	5	5
4	4	5	5
<u>6</u>	<u>6</u>	7	7
<u>6</u>	<u>6</u>	7	7

8	8	9	9
8	8	9	9
WILD CARD	WILD CARD		
WILD CARD	WILD CARD		

+10	+10	+10	+10
+20	+20	+20	+20
+30	+30	+30	+40
+40	+50	+50	+60
+70	WILD CARD	WILD CARD	WILD CARD

-10	-10	-10	-10
-20	-20	-20	-20
-30	-30	-30	-40
-40	-50	-50	-60
-70	WILD CARD	WILD CARD	WILD CARD

Name Date

Student Sheet 13

Find the Numbers

In your 1000 book, locate the right square for the following numbers and write them in your book:

45

192

850

375

799

467

903

222

631

Name Date

Student Sheet 14

Numbers in My 1000 Book

Show your 1000 book to someone at home and explain your numbering system. Ask that person to give you three numbers to find in your 1000 book. Write about the strategy you used to find each number.

1. The first number I found was ________.
   This is how I found it:

2. The second number I found was ________.
   This is how I found it:

3. The third number I found was ________.
   This is how I found it:

Name Date

Student Sheet 15

How Far to 1000?

Frogurt the Frog is jumping in your 1000 book. Frogurt jumps only 1 square at a time. Frogurt always counts his jumps, but he doesn't count the square he starts on. So if he started on 30 and took 100 jumps, he would land on 130. If he started on 50, it would take him 200 jumps to get to 250.

How many jumps will it take for Frogurt to get to 1000 from these squares in your 1000 book?

Start at 135. How many jumps to 1000?

Start at 425. How many jumps to 1000?

Start at 675. How many jumps to 1000?

Start at 815. How many jumps to 1000?

Start at 540. How many jumps to 1000?

Start at 943. How many jumps to 1000?

Start at 750. How many jumps to 1000?

Start at 756. How many jumps to 1000?

Start at 111. How many jumps to 1000?

Name Date

Student Sheet 16

What's Your Estimate?

1. Which container are you using?

2. First, just look at the container. How many do you think there are in the container?

   We think the total number is closest to (circle one of these):

   100 200 300 400 500 600 700 800 900 1000

3. Now try to figure out a better estimate without counting one by one. When you are finished, write down your new estimate and what method you used.

   Here is what we did:

   Now we think the total number is closest to (circle one of these):

   100 200 300 400 500 600 700 800 900 1000

Name Date

Related Problem Sets (page 1 of 2)

$228 + 100 =$ $200 + 300 =$ $28 + 2 =$ **$228 + 302 =$**	$450 + 100 =$ $450 + 400 =$ $400 + 400 =$ **$450 + 428 =$**
$692 - 2 =$ $692 - 300 =$ $302 + 90 =$ **$692 - 302 =$**	$989 - 100 =$ $989 - 189 =$ $989 - 200 =$ **$989 - 190 =$**

Name Date

Related Problem Sets (page 2 of 2)

300 + 100 =	600 + 300 =
342 + 100 =	678 + 100 =
342 + 150 =	678 + 300 =
350 + 150 =	675 + 325 =
342 + 149 =	**678 + 325 =**

810 – 100 =	998 – 300 =
810 – 300 =	990 – 340 =
800 – 300 =	348 + 2 =
374 + 26 =	1000 – 350 =
810 – 374 =	**998 – 348 =**

Name ______________________ Date ______

Close to 100 Score Sheet

Name ______________________

GAME 1		Score
Round 1:	___ ___ + ___ ___ = ________	________
Round 2:	___ ___ + ___ ___ = ________	________
Round 3:	___ ___ + ___ ___ = ________	________
Round 4:	___ ___ + ___ ___ = ________	________
Round 5:	___ ___ + ___ ___ = ________	________
	TOTAL SCORE	________

Name ______________________

GAME 2		Score
Round 1:	___ ___ + ___ ___ = ________	________
Round 2:	___ ___ + ___ ___ = ________	________
Round 3:	___ ___ + ___ ___ = ________	________
Round 4:	___ ___ + ___ ___ = ________	________
Round 5:	___ ___ + ___ ___ = ________	________
	TOTAL SCORE	________

Name ______________________________ Date

Student Sheet 19

Close to 1000 Score Sheet

Name ______________________________

GAME 1		Score
Round 1: ___ ___ ___ + ___ ___ ___ =	________	________
Round 2: ___ ___ ___ + ___ ___ ___ =	________	________
Round 3: ___ ___ ___ + ___ ___ ___ =	________	________
Round 4: ___ ___ ___ + ___ ___ ___ =	________	________
Round 5: ___ ___ ___ + ___ ___ ___ =	________	________
	TOTAL SCORE	________

Name ______________________________

GAME 2		Score
Round 1: ___ ___ ___ + ___ ___ ___ =	________	________
Round 2: ___ ___ ___ + ___ ___ ___ =	________	________
Round 3: ___ ___ ___ + ___ ___ ___ =	________	________
Round 4: ___ ___ ___ + ___ ___ ___ =	________	________
Round 5: ___ ___ ___ + ___ ___ ___ =	________	________
	TOTAL SCORE	________

Name Date

Student Sheet 20

Froggy Races 2

1. Three frogs had a race on a 1000 chart. Freaky Frog took 30 jumps of 25, Frogurt Frog took 30 jumps of 20, and Frannie Frog took 120 jumps of 5.

Who was ahead? ____________________

Write down or draw a picture to show two ways to solve this problem. (Use another piece of paper if you need it.)

2. How many more jumps of 25 does Freaky need to reach 1000? __________

How do you know? (Use the back of this paper to answer.)

Materials

- One deck of Numeral Cards
- Close to 100 Score Sheet for each player

Players: 1, 2, or 3

How to Play

1. Deal out six Numeral Cards to each player.
2. Use any four cards to make two numbers. For example, a 6 and a 5 could make either 56 or 65. Wild Cards can be used as any numeral. Try to make numbers that, when added, give you a total close to 100.
3. Write these numbers and their total on the Close to 100 Score Sheet. For example: $42 + 56 = 98$.
4. Find your score. Your score is the difference between your total and 100. For example, if your total is 98, your score is 2. If your total is 105, your score is 5.
5. Put the cards you used in a discard pile. Keep the two cards you didn't use for the next round.
6. For the next round, deal four new cards to each player. Make more numbers that come close to 100. When you run out of cards, mix up the discard pile and use those cards again.
7. Five rounds make one game. Total your scores for the five rounds. Lowest score wins!

Scoring Variation

Write the score with minus and plus signs to show the direction of your total away from 100. For example: If your total is 98, your score is –2. If your total is 105, your score is +5. The total of these two scores would be +3. Your goal is to get a total score for five rounds that is close to 0.

Materials

- One deck of Numeral Cards
- Close to 1000 Score Sheet for each player

Players: 2 or 3

How to Play

1. Deal out eight Numeral Cards to each player.
2. Use any six cards to make two numbers. For example, a 6, a 5, and a 2 could make 652, 625, 526, 562, 256, or 265. Wild Cards can be used as any numeral. Try to make numbers that, when added, give you a total that is close to 1000.
3. Write these numbers and their total on the Close to 1000 Score Sheet. For example: 652 + 347 = 999.
4. Find your score. Your score is the difference between your total and 1000.
5. Put the cards you used in a discard pile. Keep the two cards you didn't use for the next round.
6. For the next round, deal six new cards to each player. Make more numbers that come close to 1000. When you run out of cards, mix up the discard pile and use them again.
7. After five rounds, total your scores. Lowest score wins!

Scoring Variation

Write the score with plus and minus signs to show the direction of your total away from 1000. For example: If your total is 999, your score is –1. If your total is 1005, your score is +5. The total of these two scores would be +4. Your goal is to get a total score for five rounds that is close to 0.

1	2	3	4	5	6	7	8	9	10
11	12	13	14	15	16	17	18	19	20
21	22	23	24	25	26	27	28	29	30
31	32	33	34	35	36	37	38	39	40
41	42	43	44	45	46	47	48	49	50
51	52	53	54	55	56	57	58	59	60
61	62	63	64	65	66	67	68	69	70
71	72	73	74	75	76	77	78	79	80
81	82	83	84	85	86	87	88	89	90
91	92	93	94	95	96	97	98	99	100

BLANK 100 CHART

Practice Pages

This optional section provides homework ideas for teachers who want or need to give more homework than is assigned to accompany the activities in this unit. The problems included here provide additional practice in learning about number relationships and in solving computation and number problems. For number units, you may want to use some of these if your students need more work in these areas or if you want to assign daily homework. For other units, you can use these problems so that students can continue to work on developing number and computation sense while they are focusing on other mathematical content in class. We recommend that you introduce activities in class before assigning related problems for homework.

Ways to Count Money This type of problem is introduced in the unit *Mathematical Thinking at Grade 4*. Here, two problem sheets are provided. You can also make up other problems in this format, using numbers that are appropriate for your students. Students find two ways to solve each problem. They record their solution strategies.

Solving Problems in Two Ways Solving problems in two ways is emphasized throughout the *Investigations* fourth grade curriculum. Here, we provide four sheets of problems that students solve in two different ways. Problems may be addition, subtraction, multiplication, or division. Students record each way they solved the problem. We recommend you give students an opportunity to share a variety of strategies for solving problems before you assign this homework.

Name Date

Practice Page A

Find the total amount of money in two different ways.

8 quarters
3 pennies
9 dimes
1 nickel

Here is the first way I found the total amount of money:

Here is the second way I found the total amount of money:

Name Date

Practice Page B

Find the total amount of money in two different ways.

2 nickels
7 dimes
5 quarters
10 pennies

Here is the first way I found the total amount of money:

Here is the second way I found the total amount of money:

Name Date

Practice Page C

Solve this problem in two different ways, and write about how you solved it:

70 + 54 =

Here is the first way I solved it:

Here is the second way I solved it:

Name Date

Practice Page D

Solve this problem in two different ways, and write about how you solved it:

140 – 54 =

Here is the first way I solved it:

Here is the second way I solved it:

Name Date

Practice Page E

Solve this problem in two different ways, and write about how you solved it:

$6 \times 12 =$

Here is the first way I solved it:

Here is the second way I solved it:

Name Date

Practice Page F

Solve this problem in two different ways, and write about how you solved it:

$$36 \div 6 =$$

Here is the first way I solved it:

Here is the second way I solved it: